Science, Culture and Society
Second edition

Science, Culture and Society

Understanding Science in the 21st Century

Second edition, revised and updated

Mark Erickson

polity

First edition first published in 2005 by Polity Press
This second edition first published in 2016 by Polity Press

Polity Press
65 Bridge Street
Cambridge CB2 1UR, UK

Polity Press
350 Main Street
Malden, MA 02148, USA

ISBN-13: 978-0-7456-6224-4
ISBN-13: 978-0-7456-6225-1(pb)

A catalogue record for this book is available from the British Library.

Library of Congress Cataloging-in-Publication Data

Erickson, Mark, 1964-
  Science, culture and society : understanding science in the 21st century / Mark Erickson. -- 2nd Edition.
      pages cm
  Revised editon of the author's Science, culture and society, 2005.
  Includes bibliographical references.
   ISBN 978-0-7456-6224-4 (hardback : alk. paper) -- ISBN 978-0-7456-6225-1 (pbk. : alk. paper) 1. Science--Social aspects. I. Title.
   Q175.5.E75 2015
   306.4'5--dc23
                                    2015008230

Typeset in 10 on 11.5pt Palatino by
Servis Filmsetting Ltd, Stockport, Cheshire
Printed and bound in Great Britain by Clays Ltd, St Ives PLC

For further information on Polity, visit our website:
politybooks.com

Contents

Preface to Second Edition

The second edition of *Science, Culture and Society* has undergone substantial change. The core theoretical perspective remains the same; I am still committed to using the work of Ludwik Fleck and Ludwig Wittgenstein to make sense of science, focusing on the language, meaning construction and representations of science in the different communities of which we are members. I also use broadly the same structure as the first edition. However, much of the content is different. In particular, chapter 2 presents an account and analysis of a recent molecular microbiology experiment; comparing this to the experiments I used in the first edition reveals how far and how fast biosciences have moved in the intervening decade. Other chapters include new content in the form of more recent popular science texts, science fiction narratives and theoretical understandings of science in society. Also included in this edition is a much stronger focus on gender discrimination in science, reflecting how this topic is (finally) receiving much more attention in policy. The first edition included extensive discussion of nanotechnology to illustrate the relationship between the formal science produced by scientists and public understandings of science, but I think that the example in this edition – climate change science – is more appropriate and a much more pressing concern for all of us. My intention in producing the revised and updated edition, however, remains the same; to provide an overview and introduction to understanding science in social and cultural contexts and to show how all members of society are involved in the social construction of science.

Mark Erickson
March 2015

Acknowledgements

This book grew out of courses I taught at the University of Birmingham and, more recently, the University of Brighton and I am grateful to all my students for their comments, suggestions and insights into science in society. The book also grew out of research I carried out in a number of laboratories. I thank all those who took part in these various research projects, but in particular I want to express my gratitude and thanks to Dr Douglas Browning of the Department of Biosciences, University of Birmingham, who taught me a huge amount about scientific method and molecular microbiology and put up with my constant stream of questions as he was trying to get on with his experiments. My thanks also to Professor Steve Busby, who granted me access to his laboratories at the University of Birmingham. Professor Alistair Rae provided very helpful guidance on relativity and quantum mechanics.

It is a privilege to be part of a supportive academic community and Sara Bragg, Tom Shakespeare and Charlie Turner all deserve special thanks for offering advice and reading drafts. My colleagues and my students at the University of Brighton provide an endless source of inspiration and encouragement. At Polity, Jonathan Skerrett provided encouragement and advice.

Ljubica Erickson supported me throughout and helped with my grammar, and Fiona Sewell worked wonders on the manuscript with her excellent copy-editing (although all remaining mistakes are, of course, mine). Finally, my thanks to Sara Bragg and Milica Erickson-Bragg, who make everything worthwhile.

Introduction

This is a book about what science is, how it is made and how it is represented in society and culture. We have a range of resources available to us to make sense of science, such as journals, histories of science, popular science books and magazines, social science accounts of science and science fiction novels and films. This book examines these resources and their interconnections to help to understand what science is, how we can define science and why science matters in contemporary society. Science is fabricated from language and discourse, actions and practices, representations and material cultures. However, where many social science accounts see science as being confined to laboratories and other designated sites of scientific production, this book sees science spread through our society and culture, unfolding in multiple domains and in multiple forms. Science is a social construction, but all of society is involved in constructing science, not just scientists.

Science and technology studies (STS) has emerged as a diverse discipline that sees scientific knowledge and technological artefacts as being constructions. By this STS means that the knowledge that emerges from scientific situations – laboratories, observatories and so on – and the technologies that emerge from scientific knowledge are constructed and contingent on when and where they were made. On this view scientific knowledge is not discovered, uncovered or found, but is actively made through the actions and interactions of scientists and engineers using the resources that surround them. It therefore opposes a longstanding view of scientific knowledge as 'out there' waiting to be 'discovered' or 'uncovered' by talented individuals. From the STS perspective science and technology are social activities that reflect the social conditions of their production and the social conditions of those involved in their production. This book is, in part, an examination of the roots and current status of these ways of understanding science and technology.

However, there are a number of issues that arise from the STS position. The first is that many, or even most people who are involved in producing scientific knowledge and new technologies do not subscribe to the story that STS tells. For them, science is a progressive, neutral activity that produces true knowledge and facts about the natural world through applying a standard method. Most scientists do not think that the knowledge they produce is contingent on social factors or conditions, only that it is constrained by the limits of scientific possibility, material and technical resources or funding. The understanding of what science is from inside scientific institutions is often very different from that of STS scholars. In this book I have attempted to produce accounts of science that scientists themselves might recognize.

Secondly, understanding that science and technology are socially constructed tells us little about how and why science has a particular status in our society. In fact, it probably does the opposite. The commonly held view in Western industrial societies is that science is a form of knowledge that produces results that are more concrete, 'better' and more factual than other ways of making sense of the world. Our societies are filled with representations of science as a more precise way of understanding, of science as a solution to problems in the world, of science as a prop to shore up political ideologies, of science as creating a better future for us. The dominant story of science in society, scientism, tells us that science is a form of knowledge and a method of investigation that is separate, bounded and superior to other knowledge and ways of investigating. Social studies of science have long since debunked this myth, but it is very persistent in societal understandings and expectations of science.

A further point needs to be faced at the outset. Whilst many public images of, and attitudes towards, science are positive, a number are negative. Contemporary scientific activities that are in the public eye sometimes meet with resistance. Science's roles in genetic modification of organisms, human cloning, production of improved weaponry, or failure to warn of the dangers of food and other health scares, for instance, are obvious examples. Sometimes public representations of science confront the idea that science is always the right way forward. This contested status of scientific knowledge challenges the widely held public view of science as a 'good' thing.

These short descriptions of perspectives on science in society show that science is not a single thing but a complex social phenomenon that appears in many places in a number of different forms. By taking this as a starting point this book differs from many STS approaches. Whilst it may be the case, as STS holds, that scientific knowledge is socially constructed by those involved in its production, this book will argue that science as a whole, the science of our societies, is itself a social construct, which the whole of society is involved in creating. The process of social construction of science results not in a unitary and essential object, but in a complex, contested and contestable family-resemblance concept that holds a range of different meanings according to where it is being deployed, and by whom.

Our societies are so permeated by science, scientific knowledge and the products of scientific endeavour such as technology that all of us, at some level or other, consume representations of science and incorporate them into our everyday understandings. This happens in many different ways, through education, the media and culture, but also through scientific and medical interventions into our bodies, through working in scientific environments or being subject to scientific work regimes, through being included or excluded by formal scientific institutions, through the consumption of technologies. We constantly and continuously construct what science is in our language, actions and interactions, through deploying meanings and through having other meanings presented to us. Given the dominance of scientism in our society, we often don't have much choice in this.

This continuous social construction of science is based upon a range of resources that are available at any given time. This book investigates what these resources are and looks at the interrelations between them. A key one is what in this book is called 'formal science', the science that is done in laboratories and other scientific institutions. At the heart of formal science is the production of scientific knowledge through the work of scientists. Examining just how such knowledge emerges is instructive not least in revealing the complexity and difficulty of much formal scientific work. Formal science is an important topic for this analysis as it is the substrate that a number of other resources consume to construct their own versions of science. Yet the reverse is also the case: professional scientists working in scientific locations are constructing scientific knowledge, but are doing so with reference to the same external resources that the public are using. This book is able to look at a small range of these: histories of science, popular science texts and science fiction narratives. External accounts of science represent different understandings of what science 'is', and such representations serve both to reinforce a dominant story of science and to obscure aspects of the operation of formal science.

This book investigates the tensions between internal and external accounts, between esoteric and exoteric. The central argument of this book is that the social construction of science is a two-way street between the esoteric communities of which formal scientists are members and the exoteric, public communities to which we all belong. This frame of reference, based on the work of Ludwik Fleck and Ludwig Wittgenstein, is outlined in more detail in chapter 1. Subsequent chapters use this perspective to examine, firstly, the production and understanding of formal science knowledge in laboratories (chapter 2) and in philosophy and sociology of science (chapter 3) before going on to look at histories of science (chapter 4), scientific communities (chapter 5), popular science representations (chapter 6) and science fiction texts as a resource for the social construction of science (chapter 7). The book closes with an examination of climate change science and societal responses to this (chapter 8). We only know about climate change because of the activities of climate change scientists – knowledge emerged from their esoteric thought communities into wider, exoteric domains. This knowledge is understood and interpreted in exoteric domains in conflicting and contestable ways, and the societal response to climate change reflects both ambivalence towards science and scientism's advocacy of scientific solutions to social problems. Climate change science provides a clear example of how exoteric and esoteric communities are connected.

Many social accounts of science have argued that to understand science we need to understand society and its workings. Whilst this book supports that position it also argues the reverse: to understand society we need an understanding of science. To achieve that, we need to understand what formal science in scientific institutions is, and how scholars have made sense of it over the years. But we also have to recognize how society actively attaches meanings to science, making sense of science through using the resources at hand. We need to see the cultural resources that are used in this process and understand the relationship between science, culture and society if we are to be able to get to grips with what science is, why it is so important and why our society is inextricably implicated in it.

1

Science, Culture and Society

Someone says to me: 'Show the children a game.' I teach them gaming with dice, and the other says 'I didn't mean that sort of game.' Must the exclusion of the game with dice have come before his mind when he gave me the order?

Ludwig Wittgenstein, *Philosophical Investigations* ([1953] 1958)

What is science?

Much of this book is taken up with trying to define and describe science. Having this as a goal might seem strange – most people know what science is and use the word often in their everyday lives, scientists work away in their laboratories and produce scientific knowledge, social scientists use the word to describe a range of things that they see, and our culture is full of representations of science. Yet many scientists find it hard to explain even what their own work is and what it means to other people, let alone what science as a whole is and what it means. In contrast, social scientists and philosophers of science often can offer descriptions of what science as a whole means to us, and know what science as a project is, but find it difficult to explain the connection between this and the individual actions that take place in laboratories, or the role that science has in society. Our media and culture also 'know' what science is, and present us with images and understandings of science, but on closer inspection these representations often turn out to be crude stereotypes that reflect the prejudices and traditions of analysis that are bound up in the media community, rather than reflecting what science and scientists actually are. As for the 'lay public', when we begin to look at everyday uses of the word 'science' we can see that it comprises a range of meanings (think of the differences between 'domestic science' and 'science fiction'), few of which match up to dictionary or academic definitions of science.

As we start to define science we realize that our definitions are often in a negative form: we define science by saying what it is not, not what it actually is, yet we can see that science is very important to us. Science appears all around us, is part of our lives, but when we try to explain it to ourselves or to others we run into major problems, and often fall back on clichés. Science is a complex and complicated thing. We have problems looking at it due to its complexity, and at times it feels as if the more science that we have, the harder it is to explain and the less we feel we understand it. This loss of connection is compounded by a widespread belief that science is unitary and easily definable; we feel that we *should* be able to do better in making sense of science. Given this, our culture's frequent retreat into stereotypes and clichés in describing science is understandable.

Science in technoscientific worlds

What are we talking about when we speak of science? The word 'science' becomes attached to a great many different things in contemporary society, from laboratory practices to hair shampoos and political programmes. This profusion of attachments implies that there are problems of definition associated with the word 'science', and it may be that these are impossible to resolve. But we can also clearly discern a strong, dominant definition of science as the formal method of collecting knowledge about the natural world. We have a profusion of uses of the word 'science', but a strong, unitary definition. Why is this? There is something about the way that knowledge works in our society that makes us partition and compartmentalize experiences, objects, institutions, emotions, everything as if they were discrete entities. In the case of science, the conceptualization we have of it pushes us towards seeing science as being isolated, separate from us, and simply a neutral way to discover knowledge about nature. But when we look at it, when we think about how science slops over into so many other domains of life, we can see that this narrow compartmentalization will not do. We are probably wrong in trying to make sense of science in isolation, of trying to look at it on its own. At the very least we need to see it in context, and as soon as we start doing that we can see that one crucial thing about science is its embeddedness in society, in the social. We can argue elsewhere about the truth or falsity, the goodness or badness of science; our starting point for trying to make sense of science is to see it in its social contexts, as a human, social product.

The idea of 'technoscience', a concept first used by Gaston Bachelard in the 1950s and picked up and extended by Jean-François Lyotard in the late 1970s, is an attempt to address the need to see science in much wider contexts, and not simply as a method for producing knowledge. For Lyotard technoscience was an instrument where science ceases to be just about the generation of knowledge and becomes a tool that will produce technical innovations and, significantly, interventions into our lives (Lyotard 1991: 47). In contemporary society we tell ourselves a story that links science and technology such that they become insepa- rable. We have a difficulty in contemporary society in distinguishing between science and technology: when we look at new technologies we often think of science and when we look at science we often think of new technologies that science will provide. This is clear in much technology advertising, for example:

> *1999 The miracles of science ™*
> Today, consumers throughout the world invariably associate the red DuPont oval with leadership in innovative science-based materials. This recognition exists because DuPont manufactures quality products and protects their integrity through branding, the process of creating and disseminating a name that can be distinguished from other prod- ucts and develop customer loyalty and trademarking, which protects brand names from unauthorized and unscrupulous use. Over the years, branding and trademarking products have increased in importance as a result of DuPont's continued diversification and a steady increase in market competition. (http://www2.dupont.com/Phoenix_Heritage/ en_US/1999_b_detail.html)

This is more than a story of conflating two things, science and tech- nology, that have an elective affinity: technoscience designates a state of affairs (a time and place – Western industrial societies in the early twenty-first century) where the intellectual problems of the day become increasingly dominated by technical and mechanical consid- erations and often solutions. The recent surge in interest in epigenetics

---

Box 1.1 Genetics and epigenetics

DNA is the molecule that passes on inherited characteristics; changes in a DNA sequence will change these characteristics. This is genetic inheritance. The DNA that an organism inherits provides the code for which genes will be transcribed, and so what characteristics the organism will have. DNA is, in most organisms, a very large molecule which contains the code for many, many different things. Not all of the genome, the complete set of genes or genetic material in an organism, is transcribed; indeed, much of the human genome is described as being redundant or 'junk' DNA.

A large number of factors determine which genes will be

transcribed and when they will become active. Whilst every cell in an organism will have the same DNA, each cell will organize and transcribe the genome into accessible and closed regions using epigenetic regulators and transcription factors. This is a very complex set of interrelated processes involving DNA methylation (the usual form of which serves to prevent a gene being expressed), nucleosome remodelling, exchange of histone variants (thus changing the folding of the DNA molecule), and post-translational modification of the histones (the molecules around which the DNA is wrapped).

Experiments on laboratory animals and in vitro have shown that some epigenetic traits may be heritable, and these claims are significant. The doctrine of DNA specifies that only DNA can carry heritable traits and that only changes in the DNA sequence will change heritable traits. The implication of these more recent epigenetic claims is that changes in the environment of an organism may affect the transcription of the organism's DNA and that these epigenetic changes may be heritable by subsequent generations. 'A heritable alteration, in which the DNA sequence itself is unaltered, is called epigenetic inheritance' (Griffiths 2005: 326).

The implications could be vast, but are both contested and highly contentious. The idea that organisms adapt genetically to their environment is an old argument, dating at least back to the work of Jean-Baptiste Lamarck (1744–1829), who argued that physiological changes that an organism undergoes in its lifetime can be passed on to offspring (the inheritance of acquired characteristics); but it is a thoroughly discredited one, replaced by Charles Darwin's theory of natural selection. However, proponents of epigenetics have made bold claims concerning environmental factors changing DNA methylation; a parent's adaptation to or just exposure in an environment can be transmitted to the child. Evidence for this in humans is scant, although a study of Holocaust survivors and their children's mental health produced interesting findings (Glausiusz 2014). 'Biologists first observed this "transgenerational epigenetic inheritance" in plants. Tomatoes, for example, pass along chemical markings that control an important ripening gene. But, over the past few years, evidence has been accumulating that the phenomenon occurs in rodents and humans as well' (Hughes 2014: 22).

Language is important here; what we are talking about is really two separate concepts: the actual epigenetic processes that take place inside a cell, such as DNA methylation, and the possibility that acquired characteristics, the consequence of epigenetic activity in a cell, are heritable. Perhaps we should go further and describe these as two separate, but related, language-games (see p. 81 on Wittgenstein and language-games) where the first provides a scientistic base for the second.

is a good example of this: human problems such as crime or mental illness are traced back to technical causes which may, if epigeneticists are correct, have a technical solution.

As well as the point where science and technology become inseparable, where a particular state of affairs pertains and technical solutions become paramount, technoscience is also a context that our inquiries are located within: technoscience exceeds the sum of its parts. Donna Haraway expresses this well:

> This discourse takes shape from the material, social, and literary technologies that bind us together as entities within the region of historical hyperspace called technoscience. *Hyper* means 'over' or 'beyond', in the sense of 'overshooting' or 'extravagance'. Thus, technoscience indicates a time-space modality that is extravagant, that overshoots passages through naked or unmarked history. Technoscience extravagantly exceeds the distinction between science and technology as well as those between nature and society, subjects and objects, and the natural and artificial that structured the imaginary time called modernity. (Haraway 1997: 3)

Technoscience is thus both an object of inquiry and a context that our inquiry can be located within. Technoscience is also a language and a grammar that we are using to describe the world around us and our selves within the world. Our lives are now described by technoscientific language, our meanings are constructed around technoscientific viewpoints on the world. We cannot easily escape this frame of reference, this form of life. Judy Wajcman, taking Haraway as one of her starting points, argues that technoscience is a gendered domain (2004: 9) and that the technologies which emerge within it are both a source and a consequence of gender relations (ibid.: 107). This approach is a deliberate challenge to the idea that technology and science are neutral things that can be put to a range of uses: Wajcman's technofeminism points out that science and technology – and technoscience in particular – are gendered objects that have gendered consequences (Wajcman 2007). It isn't simply the case that men and women can and do use science and technology in different ways; it is that science and technology themselves are gendered, reflecting gender divisions and inequalities. This theme of gender division and gender inequality is one we will return to often in this book: science and technology are sexist, and science and technology must be understood as social relations as well as objects.

Given this complexity – the entanglement of technology and science, the conflation of knowledge, social relations and practices in science and technology – it will be difficult to identify a starting point for technoscience; when we talk about it we are talking about trends and tendencies, rather than facts and figures (although these, too, are important). One such trace we can identify is, ironically, in the world of artistic production; ironic because the dominant discourse about science in contemporary society is still articulating the idea of 'two cultures' (Snow 1959; Wallerstein 2006) where art and science are considered to be separate spheres, and art is removed and remote from the 'concreteness' of scientific knowledge and the tangible impact of technology.

Representing technoscience – Paolozzi and Faraday

The work of Eduardo Paolozzi (1924–2005) was inspired by technoscience – the fusing together of science and technology – from the 1950s (for examples see Paolozzi 1958 in Kirkpatrick 1970). The themes of his art and sculpture are often 'scientific' in the sense that they express the significance of scientific knowledge and scientists in our society, and he often executes works using the technology of our everyday lives – domestic appliances, engines, radios, robotic toys – and placing it in new, sometimes surreal, contexts and conjunctions. Paolozzi's commitment to modernism and to the progressive character of modernity is clear, and his art expresses a strong faith in the power of science to transform and change the world, although the results are sometimes unexpected or even surreal. Paolozzi's works show that science has creative power (in the senses both of being a product of creative processes such as human imagination, and of creating things), transformative power (changing nature and society and the self) and visual power (science looks good). His work also suggests that science is inescapable: it is an integral part of the modern world that we live in, and Paolozzi's art celebrates this, whilst at the same time showing the dangers of science and the power that science and technology have over us.

Michael Faraday (1791–1867), one of Britain's greatest scientists, made huge theoretical and practical contributions to the study and understanding of electrical phenomena in the early nineteenth century. Faraday's understanding of science was based upon the idea that the world was a structured whole, formed by continuously interacting natural agents or powers, and the task of the scientist was to discover the regular patterns in nature and to describe the laws that govern the behaviour of natural phenomena (Agassi 1971; Harré 1981: 177). In many ways, as we shall see, Faraday's understanding of science was little different from that of many people – scientists and non-scientists alike – in contemporary society: we are often told that science is a form of knowledge and set of techniques that provides a truthful account of the natural world by breaking it down into its component parts and identifying the rules and laws that govern the behaviour of such parts (this is, in truncated form, the 'standard' account of science; see box 2.1 on p. 31). Among Faraday's achievements were a series of experiments that showed that different varieties of electricity, that is, electrical phenomena produced by different means such as chemical or mechanical processes, were all manifestations of the same phenomenon. Through this Faraday unified the understanding of electricity in a way similar to Newton's unification of the laws of motion in the seventeenth century. Faraday was both an experimenter and a theorist, a point emphasized in Agassi's definitive biography (Agassi 1971), and Faraday's experimental work produced two of the most important inventions of modernity: the electrical generator and the electrical transformer. His practical and theoretical works are of huge importance: the electrification of the world and the entry of electricity and electrical modes of being are a direct consequence of Faraday's work. It is unsurprising that Paolozzi chose Faraday as a theme and subject for a sculpture (see figure 1.1).

Paolozzi completed *Faraday* in 2000 – it is a millennial piece of art, a celebration of modernity. The sculpture shows a seated figure, monumental (over 5m high), powerful, superhuman, raised above the viewer on a large pedestal. The humanoid form that is represented here is a figure that is fragmented and invaded by geometrical machine-made forms. The figure holds rods – symbols of power and law – that extrude cables that encircle the piece. These lines of power evoke those that encircle our world and encircle the self. The self is transformed through this power, changing from the organic version of a human being to a transformed 'modern' human, fractured and re-constructed by science and, by association, modernity. This is most visible in the 'cubist' way that Paolozzi represents the head of this sculpture (see figure 1.2): Paolozzi saw the cubist heads that he began to produce in the 1990s as inspired by computer graphics that had been pared down to a series of geometrical facets. He described these cubist heads as 'Mondrian Heads' 'because they reminded him of Piet Mondrian's late "boogie-woogie" paintings' (Pearson 1999: 74). The shift from an organic human form to an inorganic, machine-made humanoid form suggests that our human selves have been transformed by technology and by science, that we have become cyborgs – machine/human hybrids (see chapter 7 for more about cyborgs).

What does this sculpture say about our relationship to science? Science is an agent here, changing our lives and changing our bodies, changing society and changing nature. Science is powerful, exuding energy that can change the self and can transform the whole world. But science is also dangerous, transforming the human into the inhuman (Lyotard 1991). The understanding of science that is being represented here is not the same as that expressed by Faraday himself, nor is it the same as the dominant story of science that we are told in contemporary society.

The science of the *Faraday* sculpture, the science of today, is not the unified and essentialist science of the past, i.e. a form of science where there are clear foundations and rules that unify all scientific

Figure 1.1 Sir Eduardo Paolozzi RA, *Faraday*, 2000 (Photo by Bec Chalkley)

endeavours. Science in the past has been characterized by the unity of
the enterprise – science was seen as a combined, unified project that all
scientific knowledge was a part of, and all scientists subscribed to the
core values and goals of this project. Here, when we look at *Faraday*, we
see a void at the core of the form, the scientist, and, by implication – for
scientists are as much a part of society as the rest of us – a void at the

heart of all of us. There is no core to science: instead there is a rather glaring absence, one that is surrounded by strength and power, but an absence nonetheless. Our science today is fragmented and being constantly reconstructed by its interaction with society, economics and culture. Our science is not a pure form of knowledge and practice but is a confused and confusing set of enterprises, activities and representations that make up our technoscientific reality. There is no centre to 'science' because science has expanded far past its original boundaries and has entered and colonized realms that also have no centres.

It takes the power and insight of an artist such as Paolozzi to see this and to produce an image of this that begins to express what that absence means. We tell ourselves a story of science as unified, as powerful, as transformative and, above all, as having a continuous history and a continuous future. We experience science as an essential force, as something that has an essence, a core that is true and permanent. However, as Paolozzi's *Faraday* suggests, we need not look at science in that way. Thinking of Paolozzi's sculpture, and particularly his cubist

Figure 1.2 Detail of *Faraday* – head (Photo by Bec Chalkley)

heads, gives us another way of configuring and understanding what science is.

Paolozzi's cubist heads are made up of geometric fragments that have been recombined and reconstructed, and we can think of science as having a similar form. Science can be imagined as being a semi-opaque, three-dimensional object with many faces – a dodecahedron, for example – an object that we imagine turning around in our heads such that, as we turn it, a new face comes into view and focus.

As we turn our dodecahedron around in our mind, each face we see expresses a different aspect of science. We may start with the face that shows us the production of scientific knowledge in the lab, and we can examine this face to understand further what this process of knowledge production entails. As we look at this face of science we see that adjoining it are other faces that are impinging on the production of scientific knowledge – the history of science, the materiality of lab work, the scientists' understanding of their own project – and we will find it easier to understand the process of knowledge production by making reference to the adjoining faces. Later, we can turn the dodeca-hedron around and look at another face – perhaps the representation of science in popular culture. Again, we can examine this face to make better sense of how such representations are produced and how they are connected to, for example, the science of popular science books, the practices of laboratory workers or a common theme in science fiction. The more we look at the dodecahedron, the more we see the connections, tensions and interconnections between the different faces of science. We also begin to see that the dodecahedron is, actually, much more multi-faceted: there are far more than twelve faces to this imaginary object. And because the imaginary object is semi-opaque we also see that inside, like Paolozzi's *Faraday*, it has no core, no centre, no essence. Each facet of the object of science, like each facet of the head of *Faraday*, has a reality and an existence of its own, and is related to other facets, yet no one facet expresses all of the object, or has a necessary superiority to other facets.

When we look at an art object, say a sculpture or a painting, we often find ourselves looking at one particular feature – the eyes of a portrait, or the foreground of a landscape. That we focus on, say, the eyes of Paolozzi's *Faraday* is a feature of our culture, not a feature of the sculpture: similarly it is a feature of our culture and not an expression of some integral quality or essence of science that we look at science and describe its form of knowledge as being superior to other forms of knowledge. This need not be the case, although our investigations will show that there are features of scientific knowledge that make it differ-ent from other forms of knowledge.

We can describe science as a multi-faceted object that we can pick up, turn this way and that, peer inside and scrutinize, but science also has its own agency. With Haraway we must admit that it is no longer possible to maintain a strong separation between subject and object: things that appear to be passive and subject to external influence often turn out to be active and capable of effecting change themselves (Haraway 1997). We objectify science, but do this through our subjectivity that is itself constituted by our technoscientific lives. This means that we need to look at ourselves, our relationship to science, and how we embody science, to start to understand it. We need to see science located in its social context; not as a separate and remote object, but as something that is embedded in the world of social relations. But also we need to scrutinize science appropriately and recognize that we cannot grasp it all at once as a whole.

Two key thinkers

We can see science as a fragmented and multi-faceted object. This is a perspective that is supported by two key thinkers – Ludwig Wittgenstein and Ludwik Fleck – who offer, respectively, some tools for making sense of the grammar and language of science, and of the social relationships inside and surrounding science.

Ludwig Wittgenstein

The later philosophy of Ludwig Wittgenstein provides us with a way of understanding how a concept such as 'science', which appears to have such a tight and formal definition, can come to have so many different meanings, and how such a complex concept can occupy such a central position in social thought. Ironically, it is Wittgenstein's early philosophy that provides the opposite conception of science, the position held by logical positivists such as Rudolf Carnap and A.J. Ayer (see chapter 3), where science is seen as being a unified project with a unitary method: the most important, and best, form of knowledge existing in modern society (Ayer 1971). As Ayer notes: 'There is no field of experience which cannot, in principle, be brought under some form of scientific law, and no type of speculative knowledge about the world which it is, in principle, beyond the power of science to give' (Ayer 1971: 64).

Box 1.2 Ludwig Wittgenstein

Wittgenstein (1889–1951) was born into a wealthy Viennese family. He initially studied engineering before moving to study philosophy at Cambridge in the years before World War I. His first book,

Tractatus Logico-Philosophicus (published in 1921), was hailed as a work of genius. Wittgenstein thought that it had solved all of the current philosophical problems, and abandoned philosophy after it was published. The book presents in a rigidly logical way a series of propositions that describe the relationship of language to the world, and at the heart of the book is Wittgenstein's picture theory of meaning, which states that language consists of propositions which picture the world.

Wittgenstein realized in the late 1920s that the doctrine of the *Tractatus* was wrong and returned to philosophy, taking up a chair in philosophy at Cambridge. His second book, *Philosophical Investigations* (published posthumously in 1953), is also about the relationship of language to the world. However, in this book Wittgenstein abandons the idea that propositions have fixed meanings that can be broken down into their logical elements. He moves away from the formal analytical frame of reference and looks at how meanings become attached to words. The meanings of words were constructed through their use, and thus could not be understood when taken out of their linguistic context. Breaking sentences down into atomic elements to find meanings would never work as the meanings are attached as language is deployed. Wittgenstein called the whole situation where meanings become attached to words 'language-games'.

Wittgenstein's later philosophy radically departs from his earlier work. Where he had seen the world as being a totality of logical propositions which could be described with the regularity of scientific endeavour, in his later work Wittgenstein presented an understanding of the meanings of words as being constructed through their use – the meaning of words was contingent upon their use in everyday speech. This meant that there were no fixed meanings for words, and that meanings could shift and change according to how, where and by whom they were being used. Wittgenstein thought that some concepts in our language were 'family-resemblance concepts' in that they could not fall into simple 'true/false' bipolar distinctions (such as a colour being described as red or as not red, one or the other) and were really amalgamations of a constellation of meanings. He uses the example of 'games' to illustrate this point:

> Consider for example the proceedings that we call 'games'. I mean board-games, card-games, ball-games, Olympic-games, and so on. What is common to them all? – Don't say: 'There *must* be some-thing common, or they would not be called "games"' – but *look and see* whether there is anything common to all. – For if you look at them you will not see something that is common to *all*, but similarities, relation-ships, and a whole series of them at that. . . . Look for example at board games, with their multifarious relationships. Now pass to card-games;

here you find many correspondences with the first group, but many common features drop out, and others appear. When we pass next to ball-games, much that is common is retained, but much is lost. – Are they all 'amusing'? Compare chess with noughts and crosses. Or is there always winning and losing, or competition between players? Think of patience. . . . Think now of a game like ring-a-ring-a-roses; here is the element of amusement, but how many other characteristic features have disappeared? . . . And the result of this examination is: we see a complicated network of similarities overlapping and criss-crossing, sometimes overall similarities, sometimes similarities of detail. (Wittgenstein [1953] 1958: §66)

I would propose that we consider 'science' to be such a family-resemblance concept (Wittgenstein [1953] 1958: §67; Phillips 1977), where it is not possible to consider phenomena as either 'science' or 'not science'. Rather, we will see that much of what we formally categorize as science in contemporary society is not as 'scientific' as we thought, and much of what we think of as being unscientific actually contains elements of science. However, it isn't simply that our use of the word 'science' (or scientist, scientific, etc.) is ambiguous and open to a range of meanings. We need to recognize that, in the same way that when we deploy the word 'game' in everyday language we cannot help starting to look for game-like features, when we deploy the word 'science' we start looking for 'scientific' features. Given the open-textured definition of the words that we use, it is likely that we will find such features when we start looking. Wittgenstein uses the example of 'natural law' and the use of the term by scientists to show that simply using particular words means that we will end up carrying out certain forms of investigation, or will start to look at the world in a certain way. When we use the term 'natural law' we immediately start looking for certain things, and start thinking in a certain way:

> First of all, the idea of compulsion already lies in the word 'law'. The word 'law' suggests more than an observed regularity which we take it will go on.
> The usage of the word natural law connects, one might say, to a certain kind of fatalism. What will happen is laid down somewhere . . . if we got hold of the book in which natural laws were really laid down. The rules were laid down by a Deity – written in a book. Rules in physics are a guess: 'I suppose that is the law'. (Wittgenstein 1993: 430)

For Wittgenstein, the use by science of terms such as 'natural law' meant that science would *always* see the world as if there was a set of already written laws that it simply needed to uncover. It is not that our language 'compels' us inescapably to see things in certain ways (this would be too determinist and mechanistic for Wittgenstein), but that there is an element of compulsion there, and that it is only by great effort that we can escape from the view of the world that our language imposes upon us. We can use Wittgenstein's philosophy as a therapeutic intervention into the world of ideas and words to help us see where it is that language is leading us, and to identify alternative understandings of the world around us. Wittgenstein's philosophy helps us to see that our language, far from being a perfect tool for describing and explaining the world, is something that actually hides things from us, confuses us, leads us astray.

By applying Wittgenstein's philosophy to the case of science we can begin to see the concept of science as a complex constellation of meanings that bring together a wide range of different practices and knowledges. By comparison with Wittgenstein's example of 'game', this family-resemblance concept is much more difficult to unpack because one of the central modes of understanding the contemporary world is based upon a version of scientific knowledge, scientific method and scientific practice. There is no formal theory of games that make us look at games in our society as if there was a hierarchy of practices – with, say, chess at the top and football at the bottom. However, there is a strong perception in our society that scientific knowledge is better than other sorts of knowledge, and that scientific practice is a more reliable mode of investigation than other forms of inquiry. This story is reinforced again and again in our culture. For example, Stephen Hawking and Leonard Mlodinow's 2011 popular science book argues that the most important questions of life and how we can understand the world:

> [t]raditionally [were] questions for philosophy, but philosophy is dead. Philosophy has not kept up with modern developments in science, particularly physics. Scientists have become the bearers of the torch of discovery in our search for knowledge. (Hawking and Mlodinow 2011: 13)

In similar vein Richard Dawkins, eminent biologist and social commentator, when asked by popular science magazine *New Scientist* what would make the world better, replied: 'The world would be a better place if everybody learned to think like scientists' (*New Scientist*, 2725, 33). An op-ed piece in the next issue of the same magazine made a similar point:

> So being rational can be difficult – it often goes against our gut feelings. But if we want to stay alive, let alone make the world a better place to live in, there is no substitute for science and reason. We need to base our actions on how things really are, rather than how we would like them to be – and elect leaders who do the same. (le Page 2009)

These are strong, sweeping statements and claims, but they are so prevalent that we barely notice them even when they claim ownership of all significant forms of knowledge and meaning. The cover of *New Scientist* for 7 July 2012, reporting on the discovery of the Higgs boson (aka the 'god particle'), declared: 'Higgsteria! What the god particle means for the universe.' Philosopher of science Paul Feyerabend's double-edged statement '[s]cience seems to be about all there is' (Feyerabend 2011: 6) captures this condition well. Because of this we find it difficult to see the other uses of science, or the other locations of science, let alone other forms of knowledge and meaning, as having similar status to 'formal science' (broadly akin to the natural science taught and researched at universities) (See box 1.3).

Box 1.3 Formal science

I use the term 'formal science' in this book to denote scientific knowledge and scientific research that is formalized through institutional acceptance, principally through the formal institutions of scientific communities and scientific publications. The term 'natural science' is a charged one, implying as it does a direct relationship to 'nature'. As we will see, a lot of what scientists do is not particularly natural and doesn't actually involve objects from 'nature' (see box 3.1 on p. 73).

Using the term 'formal science' also alerts us to the problems of using the word 'nature'. The dominant story of science we tell ourselves implies that nature is something that is external to us, 'out there' and tangible as a separate realm. In contrast, and with philosopher Immanuel Kant (1724–1804), we need to see that nature is not natural, is constructed by us through our language, social relations and everyday practices. Nature is a product of human action and thought; talk of 'natural science' from this perspective is misleading.

The dominant story of science and the superiority of scientific knowledge has a long history. This story reinforces science's image of being a method of achieving truth, a discourse that is neutral in its origin and consequences. We can identify a number of different locations where this story is presented to us, and where we interact with it. We can call these different locations where stories of the world we inhabit are told and retold *thought communities*.

Ludwik Fleck

The idea of thought communities comes from the work of the Polish microbiologist and sociologist of science Ludwik Fleck (see box 1.4). A thought community is a group of individuals who think in a similar way using shared ideas, concepts and theories: they share a particular 'thought style' that is visible in the discourse of the thought community.

Box 1.4 Ludwik Fleck

Fleck (1896–1961) was a Polish microbiologist who took a keen interest in the philosophy and sociology of science. His groundbreaking book *Genesis and Development of a Scientific Fact* was published in 1935. In it, Fleck describes how scientific communities come together, and how the esoteric (internal) knowledge of a scientific community will be related to the exoteric (external) knowledge of other thought communities. In addition, Fleck offers an account of the history of science which is discontinuous, unlike the seamless progressive standard accounts.

Fleck's life is quite astonishing in other ways. By 1935 he was head of the bacteriological laboratory in the city of Lvov, but was sacked that year as part of the anti-Jewish measures taken by the Polish state. *Genesis and Development of a Scientific Fact* was published that year, but in Switzerland – as Fleck was Jewish he could not get it published in Poland or Germany due to anti-Semitic prejudice and laws. Up to 1939 Fleck survived as a private researcher, but then the Russian occupation of that part of Poland saw him made director of Lvov city microbiology laboratory, and appointed to the state medical school. The Nazi invasion of Russia in 1941 saw Fleck sacked (again) and confined to the ghetto – he became director of the bacteriological laboratory at the Jewish hospital, a post he held until December 1942.

During this period there was an outbreak of typhoid, and Fleck – without proper equipment, lab supplies or funds – managed to invent and develop a diagnostic test for typhus, allowing infected patients to be confined at a much earlier stage (thus preventing the infection spreading) and started working on a vaccine which he extracted from the urine of infected patients. He used his family to test this out, but before he could check the results in a fully scientific way, by vaccinating a large group, the Nazis liquidated the ghetto. Fleck was sent to Auschwitz, where he was forced to produce his vaccine for use by the *Wehrmacht* on the Eastern Front.

Fleck agreed to do this, and was supplied with a team and a small lab in one of the isolation huts, i.e. where all the inmates were already dying of typhoid. Fleck produced a large quantity of vaccine from the urine of German soldiers (the Nazis would not allow him to use urine from Jews to produce vaccine for their troops, of course). This 'vaccine' was inert – effectively sterile – and did not protect against typhoid. Fleck's team also produced a small quantity of real vaccine that did work – thus allowing them to convince the Nazis that what they were injecting into their soldiers was a real vaccine. This real vaccine they kept for themselves and for the inmates of

Auschwitz, as many of whom as possible were secretly inoculated. Fleck was upset that he could not keep proper scientific records of how effective these inoculations were as the Nazis would kill or relocate his research subjects frequently.

Fleck survived Auschwitz, and no doubt countless others also survived because of the vaccine he produced there. After the war Fleck was made professor of microbiology at Lublin University, and a fellow of the Polish academy of sciences in 1954. He emigrated to Israel in 1957 and died in 1961.

Fleck suggests that such thought communities are multiple, vary in size and vary in composition. For Fleck, it was important to see scientists in the context of their thought community: new scientific ideas could only 'progress' once they were accepted by the thought community that would be using them (see box 1.5).

In his major work *Genesis and Development of a Scientific Fact* ([1935] 1979) Fleck looks at how one thought community of scientists – microbiologists who were investigating syphilis – came to accept and understand a specific scientific test. Crucially, Fleck notes that the thought community of scientists would be affected and altered by the individual scientists' membership of other thought communities. For example, a scientist will be a member of her own thought community – say microbiologists – but will also be a member of a number of other thought communities: her family, maybe a political party, a workers' association, a local community, etc. Each of these thought communities has its own thought style, its own way of making sense of and under-standing the world, its own way of describing the world. The scientist cannot abandon all of these other – possibly competing – thought styles at the door to the laboratory, however much she may want to. The result is a thought community of scientists – what we may choose to call a scientific community – whose scientific knowledge, although seen as a neutral and pure product, is actually made up of knowledges that incorporate, at least to some extent, the thought styles of the communities external to that scientific community. Hence, for Fleck, societies end up with scientific knowledge that reflects the social conditions of that society as well as of our understanding of the natural world.

Box 1.5 Thought communities, thought collectives and thought styles

One of the most useful concepts that the work of Ludwik Fleck has provided is that of *Denkkollectiv*, a word variously translated as 'thought collective' or 'thought community' (in Bradley and Trenn's translation (Fleck [1935] 1979)) or as 'cognitive community' (in

Baldamus's translation (1977)). For the sake of simplicity, the term used throughout this book is 'thought community'.

Fleck defines thought community as 'a community of persons mutually exchanging ideas or maintaining intellectual interaction' (Fleck [1935] 1979: 39). This definition, whilst clearly applicable to a community of scientists working together on a specific range of problems and experiments, is also applicable to non-scientific communities. Fleck notes that 'there are structural characteristics shared by all communities of thought' (Fleck [1935] 1979: 105), and he describes two different sorts of thought communities: esoteric and exoteric. 'The general structure of a thought collective consists of both a small esoteric circle and a larger exoteric circle, each consisting of members belonging to the thought collective and forming around any work of the mind, such as a dogma of faith, a scientific idea, or an artistic musing. A thought collective consists of many such intersecting circles. Any individual may belong to several exoteric circles but probably only to a few, if any esoteric circles' (Fleck [1935] 1979: 105). Individuals can be inducted into the inner, esoteric, circle if they are already members of the exoteric circle, assuming they are sufficiently proficient in articulating the thought of the inner circle.

Fleck notes that each thought community is characterized by a particular thought style (*Denkstil*), and that the thought styles operating inside thought communities are never independent, and the members of thought communities are also not independent. Thought styles inside thought communities are related to other thought styles, firstly from the interaction between the esoteric and exoteric parts of a thought community, and secondly by benefit of the members of each thought community also being members of a number of other thought communities, each of which also has its own thought style, each of which is just as interdependent as others. We can identify thought styles from collecting and examining the discourses being used inside a thought community.

Esoteric and exoteric thought communities become useful concepts through which sense can be made of the relationship between the very specific research-oriented experimental communities and wider scientific communities. Further, these concepts allow one to see how scientific thought – on the surface neutral and independent – is actually dependent on social factors external to the workings of scientific practice.

This is a powerful challenge to our dominant story of the neutrality and truthfulness of scientific knowledge. Where our dominant story tells us that scientific knowledge is the result of a precise endeavour that produces facts about the natural world – facts that are either true or false – Fleck's understanding (and Wittgenstein's too) is quite different. Fleck categorically rejects the idea that currently recognized 'facts' are more true than facts were in the past, and proposes that we understand 'facts' as contingent on place and use in a particular location – on their genesis inside a thought community. It isn't only those inside esoteric, highly specialized scientific communities that are responsible for constructing science. Those of us in exoteric thought communities, ones external to where scientific knowledge is produced, are also involved in making up science through representing, discussing and reproducing the ideas of science.

But given that Fleck's theory is based on observations of formal scientists doing their everyday work, how is it that most scientists don't recognize the external influences that are contributing to their knowledge production? As we shall see in chapter 3, most formal scientists subscribe to the 'standard account' of scientific knowledge production which sees science as neutral, objective and free of external influence. The answer, for Fleck, is that esoteric thought communities are just that; esoteric. They become turned inwards and unreflective, failing to notice that their thought community's dominant thought style is shaping their form of knowledge, language and life (Fleck [1960] 1986; Rose 2012). In addition to this, like many people in our society they are responsible for promulgating and perpetuating the idea that scientific knowledge is somehow better than and superior to other forms of knowledge. We can call this widely spread discourse and thought style 'scientism'.

Scientism

Fleck's theory of thought communities not only provides us with a complex understanding of the contingency of scientific knowledge, but also begins to explain the expansion and colonization of science in our everyday lives. The thought communities we are a part of become infused, infected, with the ideas and ethos of science to the extent that science as a dominant or superior mode of understanding our world becomes a major thought style. We can call this infection 'scientism' – the belief or general feeling that science is the best mode of explanation for things in the world, and that science and scientific methods have a special power. Scientism delivers to us a dominant story of science in our contemporary society: a contested story that attracts dissenting voices and, at times, alienates many people from its core narrative, but a powerful and prevalent story nonetheless.

Describing scientism as a thought style distinguishes this analysis from other, earlier approaches to scientism. It has been seen as an ideology, a kind of distorted understanding of the world around us, but this conceptualization of scientism makes it appear as a totalizing and exclusive way of looking at the world only using science. That's not the case, certainly not at the moment, when we can see considerable anti-science sentiments and practices. It has also been described as a form of belief, whose central idea is that all forms of social, moral or political problems can be solved by methods similar to those used in the formal sciences. Seeing scientism as a thought style, moreover as the dominant thought style in the dominant thought community of contemporary society, allows us to see why it is that scientific knowledge is usually privileged over other forms of knowledge, and why we adopt a tech-nocratic consciousness. However, it prevents us taking this too far in the direction of totalizing or monolithic modes of perception and understanding. Mary Midgley, who defines scientism in the starkest terms as 'the ambition to take over the whole of human knowledge for physics and chemistry' (Midgley 2010: 92), explains the persuasiveness and persistence of scientism as follows:

> What we 'lay people' (as we are significantly called) mostly notice about the sciences is simply their power. Technology impresses us so deeply that we are not much surprised by the claim that scientific methods ought to be extended to cover the rest of our thought. (Midgley 2001: 59)

We ask for more science to be provided in more and more areas of our lives, and we want science to do more for us. This is often inappropriate:

> Irrelevant notions about how to make thought 'hard' and scientific by imitating physical science still constantly distort the social sciences and many other areas of our thought, notably psychiatry. Though the entire enterprise of making all our thought on human affairs conform to physical patterns has never been at all successful, the idea that we must somehow do this impossible thing still haunts us. Many people find the prospect of abandoning that attempt unbearable. (Midgley 2001: 150)

Midgley is describing the tendency towards scientism in the social sciences, something which has been at the heart of their project since their inception in the early nineteenth century. Early sociologists Henri de Saint-Simon and Auguste Comte, their followers such as Karl Marx and Herbert Spencer, and later generations have all made strong claims to their theory and/or discipline being a science with scientific methods and providing scientific knowledge. This tendency is not confined to the social sciences; many types of formal knowledge production make similar claims. In philosophy, Wittgenstein was particularly scathing of those tempted by scientism:

> Our craving for generality has another main source: our preoccupation with the method of science. I mean the method of reducing the explana-tion of natural phenomena to the smallest possible number of primitive

laws; and, in mathematics, of unifying the treatment of different topics by using a generalization. Philosophers constantly see the method of science before their eyes, and are irresistibly tempted to ask and answer the question in the way that science does. This tendency is the real source of metaphysics, and leads the philosopher into complete darkness. (Wittgenstein 1969: 18)

Outside knowledge production, we look to science for explanations of all aspects of the world around us and increasingly expect science to provide solutions to all social, technical and environmental problems. Consider societal responses to climate change: technology is seen as a causal factor through its proliferation in industrial societies and science is called upon to rectify this through various schemes such as geoengineering or carbon capture and storage (see chapter 8). In the case of technoscientific solutions to climate change there is a wide range of reasons as to why individuals and states make pleas for technical solutions to what is a social problem (increasing greenhouse gas emissions due to over-consumption in our lifestyles), and we don't have space to list them here. However, the impetus for asking for technoscientific solutions comes from the prevalence of scientism in the largest thought communities in our society. This point is made forcibly by one of the key voices contributing to early social constructionist accounts of science in society, philosopher Paul Feyerabend. His 'anarchist' theory of knowledge sought to reveal the 'myths' and 'fairy-tales' that attended science:

The image of 20th-century science in the minds of scientists and laymen is determined by technological miracles such as colour television, the moon shots, the infra-red oven, as well as by a somewhat vague but still quite influential rumour, or fairy-tale, concerning the manner in which these miracles are produced. According to the fairy-tale the success of science is the result of a subtle, but carefully balanced combination of inventiveness and control. Scientists have *ideas*. And they have special *methods* for improving ideas. The theories of science have passed the test of method. They give a better account of the world than ideas which have not passed the test. The fairy-tale explains why modern society treats science in a special way and why it grants it privileges not enjoyed by other institutions. (Feyerabend 1978a: 300)

Yet we can also see a number of challenges to the hegemony of science and scientistic thinking in our technoscientific society. Ironically, given the dire predictions made by thinkers such as Max Weber and Herbert Marcuse early in the twentieth century, we are not seeing the increasing rationalization of all aspects of our lives (although that is certainly happening to large parts of our lives). We are seeing a decrease in rationalization in, for example, the expansion of new age thought, the increase in opposition to 'scientific' interventions in our health and food, and the increasing rejection by young people of science subjects in education. Our thought communities, it would appear, are becoming more profuse, more diverse and more distinctive from one another, and the grip of the ideas of science inside them would appear to be loosening. This is coupled to another significant trend: that as we ask more of science, and more science is done we understand it less, and the less trust we have in it.

Scientism colonizes our everyday lives in a way that makes us all responsible for creating the dominant story of science in society, where science is seen as being separate, different, superior. We tell ourselves that there is society and then there is science. We can see plenty of opposition to science in anti-science trends such as the US creationist or intelligent design movement, new age mysticism, religious fundamentalism promoting reactionary attitudes towards new scientific breakthroughs such as stem cell research and genetic modification of organisms, climate change denial. Yet despite this we are still surrounded by scientism with its message of 'science first'. Why? Because we are surrounded by science; science is everywhere.

Science is everywhere

Our technoscientific world is suffused with science: scientific knowledge is imperative for the maintenance of modern lifestyles; our understanding of the world often relies on modes of thinking that, at the very least, owe a debt to the tradition of scientific investigation; our culture – popular, high, underground – relies on science and technology for the material means of production and reproduction, and science and technology are frequent themes for the content of science – from science fiction to the subject matter for the sculptures of Paolozzi. Science is a central tool in the search for power, allowing us to control our environment, and legitimating the forms of domination that exist in our industrialized societies. Science is a central component of the operation of capitalist enterprises, providing the knowledge that enables the development of new goods and services, and providing wealth for those who control it.

Our understanding of science will inevitably be affected by the stories we have already been told about it, the most dominant one being scientism. Scientism presents us with a picture of science as a neutral, objective set of tools that are the best way of describing the world around us. Whilst it is appropriate to argue that the best mode of understanding basic biochemical processes, for example, should be based on the methods, tools and understandings of formal science, many would challenge the imposition of scientific thought elsewhere.

We can also note that scientism, for all its prevalence, actually doesn't describe much of our world – our personal experiences are often at odds with the claims made by scientism. This may be due to our membership of thought communities that are becoming less easily penetrable by scientism, and our membership of various thought communities will also affect our understanding of science in society. Those who are part of a community that has suffered badly from the imposition and failure of technologies in their lives (such as the residents of the communities adjoining the Fukushima nuclear power stations), or from the domination by a scientific elite (for example, victims of unethical experiments carried out by pharmaceutical companies), will have perspectives concerning the role of science in society that are at odds with those who are members of communities and societies that are 'winning' because of the role of science in their lives.

This means that we need to understand science in a range of contexts: our scientistic, technoscientific frame of reference may not always be appropriate. And, further, we must also recognize that our frame of reference prioritizes and fetishizes science in certain locations: we 'know' where real science is – the laboratory, the textbook, the documentary – and we 'know' where non-science is – the TV screen, the science fiction book, the art gallery. We need to challenge this way of looking at science to understand what science really means to us and where it is located.

When we start to do this we begin to see that science is located almost everywhere in our society: obviously in some places – the laboratory, the school – but also in culture, in our everyday descriptions of the world, in the operation of our economic systems. Each of these sites tells us something different about science, sometimes reinforcing and sometimes contradicting our dominant story of science. Examining science as a complex object that has multiple, interdependent locations leads us to a number of key conclusions:

- Science is multi-faceted.
- Science is a complex and contestable concept.
- There is no essential core to what we call 'science' in contemporary society. Each facet of science leads us to an understanding of a different reality of science, and each has some validity.
- Placing a hierarchical order on top of these multiple realities is a social construct, not a necessary feature of science itself.

As we begin to explore science, each facet that we see brings new and different meanings with it. Our understanding of science, technology and technoscience – to name but three important things – changes according to which face of science is confronting us. In doing this we should reflect also that we are ourselves responsible for bringing some of each of these meanings and themes with us – all of us are actively involved in constructing the meaning of science and the contexts within which it is located.

We will start our exploration of the different facets and locations of science by considering what many of us would identify as being 'real' science: the laboratory work from which the majority of formal scientific knowledge emerges. Although the laboratory has become a metaphor for control and simplification in our society, what we find when we enter the actual laboratory worlds of scientists is incredible complexity, and a certain amount of chaos.

2

In the Laboratory

> *Photographs* of Einstein show him standing next to a blackboard covered
> with mathematical signs of obvious complexity; but *cartoons* of Einstein
> show him chalk still in hand, and having just written on an empty
> blackboard, as if without preparation, the magic formula of the world.
> Roland Barthes, 'The Brain of Einstein' (1993)
> Reprinted by permission of Hill and Wang, a division of Farrar, Straus
> and Giroux, LLC, and The Random House Group Limited.

In this chapter the word 'laboratory' is used as shorthand for any location where scientists are involved in the production of formal scientific
knowledge – such locations include observatories, field stations and
also computer-generated experimental areas. Although science perme-
ates all of society and is visible in multiple locations, the laboratory
remains the prime site for the enactment and execution of formal
science. The scientific knowledge that emerges from these workplaces
provides a significant resource for wider societal understandings of
science, and by extension technology. Frequently, this knowledge is
a result of the execution of experiments, and in this chapter we will
examine one experiment in detail to introduce and examine the theme
of formal scientific method.

Although the focus here is on laboratories and scientific workers in
UK universities, scientific work takes place in many locations, using a
very wide range of tools, techniques and understandings, and is done
by a very large number of people. It is difficult to come up with a precise
figure of how many people are employed in science, engineering and
technology (SET) occupations in the UK, but recent estimates propose
about 5.5m, of whom 674,000 (13.3 per cent) are women (Kirkup et al.
2010: 75). These SET workers are distributed across SET and non-SET
industries: non-SET industries are employing about 1.3m SET workers
(ibid.). Looking at university and research centre environments provides some more precision. The UK Resource Centre (UKRC) report

Table 2.1 STEM academic staff by gender, grade and mode of employment at UK HE institutions, 2007/8

Gender	Full-time researcher	Part-time researcher	Full-time lecturer	Part-time lecturer	Full-time senior researcher / lecturer	Part-time senior researcher / lecturer	Full-time professor	Part-time professor
Women	5,375	1,045	2,065	1,410	1,790	355	540	55
Men	12,355	800	5,845	2,160	8,010	465	5,265	630
Total	17,730	1,845	7,910	3,570	9,800	820	5,805	685

Source: Kirkup et al. (2010: 65, Table 4.1.1).

into gender and science, technology, engineering and mathematics (STEM) occupations collected figures from UK higher education (HE) institutions, and these are a good guide to the number of people actually involved in making formal scientific knowledge.

Table 2.1 shows over 48,000 staff in total. The gender disparities in grade are quite stark: women are much more likely to be in part-time jobs, are more likely to be in lower-grade jobs, and are less likely to achieve the highest grades. There is a 'leaky pipeline' here (see chapter 5): women passing through the career progression route in STEM subjects are much less likely to reach the end point of possible career trajectories (only 9.3 per cent of full-time professors in STEM departments are women, although 33 per cent of UK university graduates in STEM subjects are women).

Looking at how science is done starts with making sense of laboratory work, and this will provide some familiarity with the language and meaning construction that is occurring inside formal scientific workplaces. The following example of a series of molecular biology experiments that lead to a single paper published in an academic journal may present a challenge to readers who are not familiar with this form of work, but it is worth persisting with this illustration of what is, at root, the basic substance of the 'science' that is so prevalent in society and culture. The experiment discussed in this chapter is not necessarily typical or representative of formal scientific experiments in general – finding a typical experiment is not possible given the range and variety of experimental techniques and scientific disciplines – but it does illustrate how scientific method is put into practice by scientific researchers.

Box 2.1 Scientific knowledge and scientific method

The standard account of science, and the account that is still largely held by most scientists and by most lay observers of formal science, is based on the idea of scientific method being a special way of discovering facts, where facts are individual packets of truth that describe the natural world, or nature. Science, by this account, has a privileged relationship to nature: it can provide us with truthful accounts of what nature really is. Not only that, but science can, and does, act as an exemplar for other forms of human inquiry, such as investigation into the form and structure of the social world or of individual mental states.

Schematically the standard account of formal science work looks like this:

• Scientific theories generate hypotheses which can be translated into specific testable questions.

- Experiments are designed which will provide answers, in the form of facts, to those questions.
- Facts are compiled to confirm or deny the validity of the theory.
- Science thus proceeds by the operation of theory in close relationship to the facts generated by scientific endeavours using scientific method.

This position, broadly speaking, is congruent with what is called positivism (Chalmers 2013: ch. 1).

This idea would appear to distinguish scientific from other kinds of knowledge in a quite marked way. Whilst we, in our everyday lives, may know many things, there are few things that we can, or need to, prove. Science, by the standard account, proceeds through proving things beyond reasonable doubt, and by compiling facts and fact-like statements and through extending theories of the natural world.

We will look in more detail at the standard account of science in chapter 3.

The experiment described in what follows shows a number of things. It illustrates the complexity of this kind of scientific work, and the degree of detail that experimenters routinely work with. It also shows that the connections between the 'big problems' that science addresses and the actual practical activity of many scientists are loose and stretched. The experiment reveals as well the site of the production of formal scientific knowledge, and the site of the expression and application of formal scientific method.

Formal scientific knowledge can look as though it is written in a different language, for example 'Requirement for Two Copies of RNA Polymerase Alpha Subunit C-Terminal Domain for Synergistic Transcription Activation at Complex Bacterial Promoters' or 'Mutational and Topological Analysis of the *Escherichia coli* BamA Protein' (both of which are titles of molecular biology papers published in academic journals (Lloyd et al. 2002; Browning et al. 2013a)). It isn't, and Wittgenstein's work is particularly helpful in reminding us that there are no private languages, nor are there languages that do not rely on society for their meaning construction (Wittgenstein [1953] 1958: §293; Hacker 1986: ch. 9). But it may be helpful to consider formal scientific papers to be written in a 'dialect'; to understand what is happening in these papers, a process of 'translation' is necessary, moving from the tightly controlled formal scientific dialect to looser everyday language, from esoteric to exoteric. Ironically, this process is in many ways the opposite of the process that leads to the construction of the scientific paper where the pragmatic methods of a team of experimenters, with associated uncertainty, informal understandings and 'making do', are turned into the formal scientific knowledge and discourse of a science journal article, moving from exoteric to esoteric.

Throughout the description of this example a process of translation takes place. The experiments carried out in this laboratory are described by the researchers doing the work, and by the formal record of the work (i.e. scientific papers) in the 'dialect' that relies upon a large number of tightly defined, but esoteric, terms such as 'polymerase', 'molecule', 'epitope'. To understand fully what is happening in laboratory situations we need to translate these terms into a language that is more familiar to us, the 'lay' public. In this process there is an inevitable change in meaning – we are moving from one thought community, one language-game, to another. The context of the words and terms being used is dramatically different: in the laboratory situation, or in the scientific paper, the words being used must have very precise and tightly defined meanings to ensure clarity of expression and comprehensibility across the esoteric scientific community that the discourse is being addressed to. Readers of the scientific paper who are inside this community will attach meanings to terms such as 'epitope' and 'polymerase' according to their past experience, their training, their practical activities of using the things with those names. They understand the esoteric terminology and language, and understand it in pretty much the same way as the authors of the paper. We know this because we can observe this esoteric community talking to itself in the laboratory and at gatherings such as academic conferences. For outsiders these terms will retain their esoteric character and it may be difficult to attach meanings to these words in ways similar to insiders. This process of translation from scientific dialect to everyday language can never be perfect, and no attempt is made here to try and perfect this process. What may happen is that readers will become more familiar with some of the discursive constructions taking place inside this scientific workplace, and will see how an esoteric thought community structures its activities around a shared set of practices, attitudes and linguistic descriptions of the(ir) world.

Box 2.2 Experiments

Experiments are a key component of the formal scientific method. In formal science, this method begins with the collection of observed facts, and the construction of a model, a hypothesis, to explain the observed facts. The hypothesis may or may not be correct, and needs to be tested. Experiments are designed where the parameters contained in the hypothesis can be controlled and manipulated such that measurable results can be collected. The results will, ideally, confirm or deny the hypothesis. Assuming the hypothesis is confirmed by experiment it can begin to change its status. Rather than just being an educated guess that has some evidence to back it up, the hypothesis will be tested rigorously in a number of ways. It will begin to take on the status of a scientific theory.

This realist outline of scientific method is widely held by those involved in formal scientific experimental work (although as Steve Fuller notes many scientists consider it to be a 'caricature' (Fuller 1997: 12)) but it has come to be challenged strongly by many sociologists and philosophers of science.

In the molecular biology laboratory: investigating the structure and function of a protein

Before we look in detail at how this experiment is done we need to think about why it is done, and therefore we need some background and context. The basic principles of molecular biology that inform this experiment are shared across a great many current topics in biology such as genetic engineering, genomics and cloning. The specialist knowledge that is needed to carry out this experiment is the knowledge that brings together a thought community that we can call the molecular biology thought community. All members of this thought community share an understanding of how biological systems work; all living cells operate according to similar principles. Each cell contains DNA *(deoxyribose nucleic acid)*, a molecule that has the ability to replicate itself, thus giving it the ability to transmit biological information to successors, and which contains the genetic code for the production of proteins. All life processes in all living cells are carried out by proteins, which are complex molecules, and understanding how proteins are made inside cells is central to understanding any biochemical system. One of the discoverers of the structure of DNA, Francis Crick, predicted that information flow from nucleic acid to protein is not reversible, and he referred to this as the 'Central Dogma' of molecular biology (Horton 2002: 5).

In schematic form, the production of proteins in cells is a simple two-step process (see figure 2.1). The first step is called transcription: a single stranded transcript of the double-stranded DNA molecule is made by an enzyme (enzymes are also proteins) called RNA polymerase (RNAP). It does this by 'unzipping' the double-stranded DNA molecule, 'reading' the code that DNA contains, and transcribing this onto a new, single-stranded molecule called RNA (ribonucleic acid). This single-stranded molecule contains the code for a sequence of amino acids (simple nitrogen–carbon compounds) which will make a

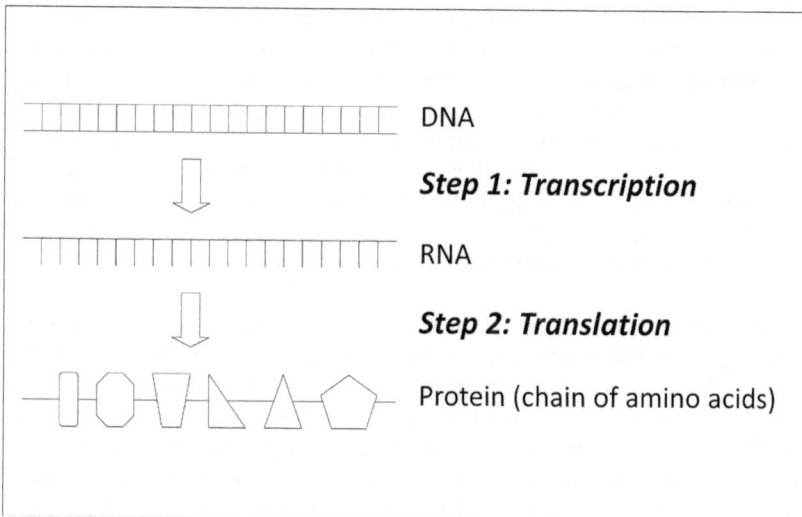

Figure 2.1 Schematic stages in protein production

protein molecule, and the RNA is subsequently translated, in step 2, into a protein.

The work of the molecular biology thought community is built on this theory; it acts as a 'paradigm' that informs and directs the thought and actions of this thought community, and others in the wider field of biology. But this paradigm is not confined to being deployed in esoteric thought communities; it is widely distributed into the exoteric in the form of school textbooks, science fiction, TV documentaries and so on. The paradigm (see chapters 3 and 4) is widely known and acknowledged in society.

However, inside the molecular biology thought community we can find more esoteric, smaller thought communities that come together around more specialized knowledge. So, where many people in exoteric thought communities understand the basic principles of how DNA operates, all molecular biologists understand the general principles of cell structure and function. But inside this grouping the eukaryotic molecular biology thought community focus on eukaryotic cells (these are cells with a highly organized nucleus, e.g. mammalian cells), the prokaryotic molecular biology thought community focus on prokaryotic cells (these are cells where there is no membrane separating the DNA-containing area from the rest of the cell, e.g. bacterial cells), virologists focus on viruses, and so on. Being a part of an esoteric thought community means sharing a base of knowledge but also sharing a style of thought; members of scientific communities, particularly very small communities, work closely with one another, either inside shared laboratories – where work is often intense and teams will spend a lot of time in each other's company – or in a range of communications – face-to-face at specialist conferences, using email, Twitter, blogs and Skype, or using 'traditional' mail. Their communications with one another rely on seeing the world in similar ways. This becomes particularly clear when one examines a formal scientific paper, as we shall do in what follows; without specialist knowledge it is difficult to make sense of what is happening, and difficult to understand the formal scientific language (see box 2.4 on p. 56).

We will look in some detail at a single formal science paper, considering why the work was done, how it was done and what the results show. More accurately, we are looking at a series of three linked experiments, carried out by a team of thirteen researchers at the University of Birmingham, that are brought together to investigate the structure and function of a protein. The experiments share an aim, and are brought together as a report in a single publication titled 'Mutational and Topological Analysis of the *Escherichia coli* BamA Protein' (Browning et al. 2013a). *PLOS ONE*, where the paper was published, is an open-access journal that publishes peer-reviewed scientific papers (see box 2.3). The paper's title identifies the method used to attack the research question: what is the structure and function of this protein? Mutational analysis is a method whereby inferences are drawn about the structure of the protein by creating a variety of mutant bacteria constructs (mutagenesis, or mutagenizing the bacteria) and subsequently measuring the functionality of the mutants. Topological analysis involves working out the protein structure from looking at the DNA sequences that code the protein and then modelling the structure and shape (topography) of the protein. The linked experiments are directed towards investigating just one part, BamA, of *E. coli's* complex β-barrel assembly machine (BAM).

Box 2.3 Peer review and open access

Securing a publication in a reputable scientific journal is a significant goal for most academic scientists: the very act of publishing verifies the quality and 'correctness' of the research. This ascription of status to formal scientific knowledge is made possible by the peer-review process, whereby other researchers who are familiar with the general field of the research described in a paper are asked to comment on the merits and demerits of the paper on behalf of the journal. Peer reviewing in the natural sciences is, in general, a one-way anonymous process: the reviewer will know the identity of the author of a paper, but the author will not know the identity of the reviewer (however, it is possible they can guess the identity). Peer review is at the heart of the process of the validation of scientific knowledge, and is also a mainstay of the process of awarding research grants: it is seen as being a way of guaranteeing the quality of scientific research.

However, as Steve Fuller points out, there are problems with the process of peer review. Given the complexity of the work carried out in many experiments, and the very high degree of specialization, it is unlikely that the peer reviewer will have extensive knowledge of the material presented in the paper, and almost inconceivable that the peer reviewer will have the time or resources to replicate the experiments being described. At best, the reviewer will be able to judge from their experience that the work described is plausible, is competently executed and produces results that are consistent with others. Fuller describes the peer-review process in terms of a collective insurance policy that a scientific community takes out (Fuller 2000: 65). But the peer-review process systematically reproduces structural problems built into scientific communities:

> [B]ecause each scientific speciality is dominated by a few gatekeepers who pass judgement on everyone else in the field, failure to appease these 'peers' can be disastrous, much like the failure to pay money to the local mafia boss. Not surprisingly, scientists tend to underplay their own originality and overplay that of their significant colleagues. (Fuller 1997: 65)

Peer review is still at the heart of formal science journals despite significant recent changes to publication formats. Following a groundswell of public opinion that publicly funded science research should be made available to the public, coupled to the rise of the world wide web in the 1990s, formal science journals are increasingly adopting an open-access format. Open access comes in a variety of forms, but the basic principle is that scientific research published in academic journals will be immediately, or within a fairly short period of time, available to anyone who wants to read it. Beneficiaries of open access are, primarily, other researchers, who get immediate and unrestricted access to their colleagues' research; in Global South countries the removal of access costs to journals is particularly significant. The general public also benefit in that individuals can access journals directly via the internet and the general dissemination of scientific knowledge to a wider audience is a public good. However, formal science journals still need funding to run, and to make a profit in many cases, and the cost of the shift to open-access publication is significant. Open access shifts the burden of payment for scientific knowledge from the reader (who used to pay via journal subscription fees) to the author, who now has to pay to have their article published. The fees can be quite high; at the time of writing (March 2015), for example, the highly rated journal *Genes & Development*, published by the commercial Cold Spring Harbor Laboratory Press, charges $2,000 to authors who want their work to be available via open access (http://genesdev.cshlp. org/site/misc/open_access.xhtml), while PLOS (Public Library of Science) – a group of non-profit journals that have always been fully open-access – charges a publication fee of between $1,350 and $2,900 per paper, depending on discipline:

> To provide Open Access, PLOS uses a business model to offset expenses – including those of peer review management, journal production and online hosting and archiving – by charging a publication fee to the authors, institutions or funders for each article published. (http://www.plos.org/publications/publication-fees)

Research funders now have to factor in publication costs when funding research projects, and as much UK research is funded from general taxation, the shift to open-access publication is being paid for by the general public. A further concern is that if publishers are generating income by accepting papers, there will be an incentive to accept more papers and there could be a drop in standards.

The experiments we will look at here come from a team working in the prokaryotic molecular biology thought community. But just identifying the origin cannot explain why these experiments were done; after all, there are millions of different types of bacteria that can be examined, and each cell exhibits an astonishing complexity. The experiments considered here look at one protein in one organism. Why?

Bacteria are a domain of single-celled organisms that are diverse and incredibly abundant: a gram of soil contains typically 40 million bacteria and the human gut alone contains 10^{14} bacterial cells, ten times the number of human cells in the body. Despite being very small (typically a few micrometers in length) they exhibit complex cellular structure. Although electron microscopes can reveal some larger structures (flagellae, for example), the detail of how these structures are constructed is hidden from us. Molecular biologists know from previous work over the last century or so that these structures are almost always made from proteins.

Finding out how bacteria work – what their structure is and how these structures produce varied functionality – is very important. Some bacteria, such as the *Escherichia coli* (*E. coli*) in our gut, are essential for human health, but many other bacteria are responsible for infections and diseases that can blight our lives, causing misery and death in millions of people, and *E. coli* in the wrong place can also cause severe illness. Antibiotic resistance is on the rise and many consider this to be a major threat facing humanity:

> CDC [Centers for Disease Control and Prevention] estimates that in the United States, more than two million people are sickened every year with antibiotic-resistant infections, with at least 23,000 dying as a result. The estimates are based on conservative assumptions and are likely minimum estimates. They are the best approximations that can be derived from currently available data. (Centers for Disease Control and Prevention 2013: 6)

Losing the ability to fight and treat bacterial infections would be a catastrophe for humankind, and extensive efforts are being made to find new antibacterial agents and therapies.

Existing antibiotics have only been with us for some eighty years, and many of these were discovered accidentally through trial and error rather than by design. Current efforts to find new antibiotics rest on finding ways that stop or even kill bacteria whilst not harming host organisms, and many focus on disrupting the ways that bacteria work. This is by no means simple: apart from anything else, our understanding of the structure and function of bacterial cells, whilst being the subject of great efforts in recent decades, is still rudimentary simply due to the sheer complexity and diversity of these organisms. Although we know that bacteria operate in ways that are largely similar to mammalian cells, using DNA transcription to produce the RNA and then the proteins they need to function and reproduce, we cannot simply 'block' this process, as this would inevitably block our own cells from working and reproducing. More subtle methods are required, and this means looking in great detail at the specific features of bacterial cell structure and function.

But if you can't see the structures, how can they be examined? Biochemistry, genetics and molecular microbiology can all play a role in this. The biochemical processes taking place can be identified and, by inference, it is possible to work out what cellular structures are involved and what they are doing. Genetic engineering can create new strains of bacteria that have subtly altered structures; comparing these strains allows inferences to be drawn as to what is happening inside the cells. Molecular microbiology can model structures and from these infer function. These are separate, but linked, approaches, all relying on the underlying paradigm of the biological sciences – the structure and operation of DNA coding and transcription, where the molecule DNA holds a code for a protein that is transcribed into RNA and this is subsequently translated into a protein molecule.

One thing that all bacteria must be able to do is transport chemicals across their cell membranes. They do this via structures embedded in their membranes (complex proteins that act as portals); there are many of these and they all have specific and largely separate functions. Finding out how these structures work and what their component parts are may provide insights into how to stop them working properly, thus providing routes to new antibiotics. In this brief account we can see that the central paradigm of molecular biology informs the understanding of individual processes inside cells, and also suggests a method for understanding structure and function in more detail (using mutation and genetic engineering). But the core theories of molecular biology cannot explain why the experiments are taking place; that is a social process and in this case the reasons are twofold. The first is that the team of experimenters have been working in a similar field for a long time; they are working out very specific problems and gradually coming to a better understanding of a complex system. The second is that they hope that a larger problem – antibiotic resistance – may be addressed, at least in part, by their work.

These, then, are the prerequisites for the experiments: a shared understanding of the core theories and principles of molecular biology (and these are revealed through the citations in the final formal scientific paper that is the outcome of the experiments), and a shared stock of known, tried and tested procedures which are used in the experiment (and these are revealed through the paper's 'methods and materials' section).

The experiment examines the structure and function of a single protein, BamA β-barrel (pronounced 'bam A beta barrel'), which is one of these transport mechanisms. It is an essential component of many prokaryotic cells and, specific to this experiment, the particular version of this protein is part of the bacterium *E. coli*. Given the prevalence of β-barrel proteins in bacterial cells, finding out the structure and function of this protein will be useful. 'β-barrel membrane proteins are essential for nutrient import, signalling, motility and survival' (Noinaj et al. 2013: 385). *E. coli's* BAM (β-barrel assembly machine) is responsible for transporting proteins made inside the bacteria to the outside of the bacteria and then embedding them in the bacteria's outer membrane (see figure 2.2).

The entire BAM structure is made up of five parts: the protein BamA (subject of these experiments) that is embedded in the bacteria's outer membrane, and four associated lipoproteins (BamB, BamC, BamD, BamE) that reside in the bacteria's periplasm. This multi-protein BAM is responsible for folding and embedding β-barrel proteins into the outer membrane of the bacteria.

BamA is an essential protein in *E. coli*. 'In spite of its importance, no study has systematically mutagenized BamA to determine regions of both structural and functional importance' (Browning et al. 2013a: 3). The team set themselves two, linked, research questions:

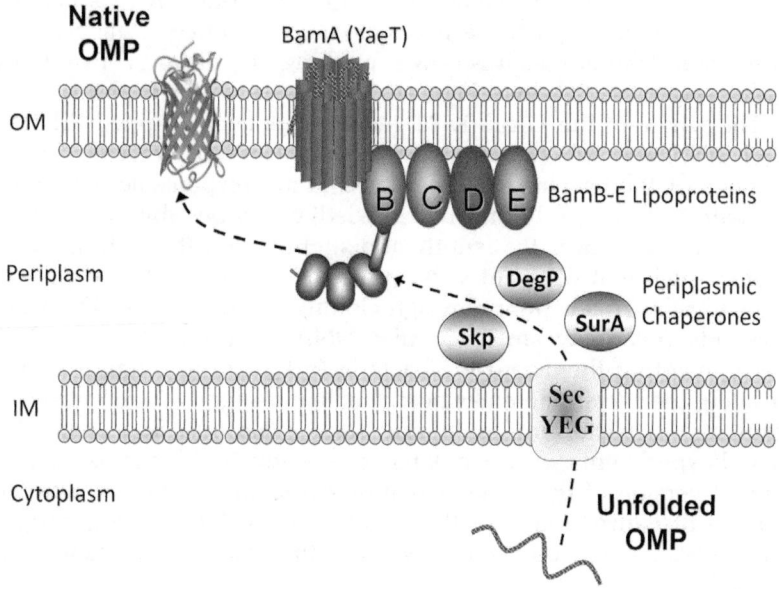

Source: Courtesy of Douglas Browning.

Figure 2.2 How BAM works (OM = outer membrane, IM = inner membrane, OMP = outer membrane protein)

- What is the organization of the BamA complex and which regions are necessary for it to function?
- Are the predictions we have made for the topology of BamA correct?

Experimental design

The basic principle at work here is to make different versions of BamA and to see what mutations do to the operation of the whole BAM complex. Whilst the experimental technique is very complex and convoluted, the basic idea is simple: the team are going to break bits off, cut bits out, or add in new bits to BamA and will then look and see if the new versions work, and if they do how well. This approach is similar to how a mechanic might attack a difficult problem in a car: trial and error, taking things out and putting them back to see where a problem lies, or how an existing system can be made more or less efficient.

The team decided to use a fairly stable strain of *E. coli*, K-12, to work on: the stability of K-12 is helpful and a lot is known about it. It was first isolated in 1922 and is one of the most common laboratory strains

of *E. coli* in use. There are some disadvantages, though; K-12 is a laboratory strain which is particularly easy to work on, but over ninety years
of living in laboratories it is now something of a distant cousin to the
'wild type' *E. coli* that lives in our intestines (Browning et al. 2013b).

The whole of the *E. coli* K-12 genome has been mapped (Blattner
et al. 1997) and subsequent analysis of this code has revealed which
sections of the bacteria's DNA molecule are responsible for coding
the BamA structure. From this knowledge it is possible to construct
new DNA sequences that subtly and slightly alter the coding for the
protein such that different versions of the protein will be expressed
in new cells. This is done through cloning small pieces of DNA into
plasmids (relatively small, circular DNA molecules (Horton 2002:
739)), inserting them into the bacteria to transform them, and then
growing the bacteria to make new, viable but slightly different strains
(mutants); a bacterium containing such a plasmid is a clone. The
linked experiments we are looking at here rely on this approach: each
altered version of the bacteria will be expressing a slightly different
version of BamA. Many of the alterations involve deleting parts of
the protein, while others involve inserting different elements (see
figure 2.3).

These mutant strains can be examined for their biochemical function
to see what they can do, at what rate and with what co-factors, and a
range of tests will be used. The results of these experiments allow the
researchers to draw inferences about how the original protein works.
This is done by comparing the mutant against the original and observing differences. Throughout the process of linked experiments, checks
are made on the results by triangulation: inferences drawn from one set
of tests are compared to inferences drawn from different tests to ensure
that the results have validity.

Method

The invention and development of the polymerase chain reaction
machine (PCR machine, also known as the thermal cycler) revolution-
ized molecular biology.

A PCR machine (see figure 2.4) can manufacture multiple copies of
chains or circles of DNA (plasmids) according to the coding sequence
it is given. Normally they are located inside a cell separately from the
main DNA of the bacteria (see figure 2.5).

Plasmids are quite remarkable things. They code for specific proteins to be made inside a cell using the existing cell machinery (e.g.
RNA polymerase and the cell's ribosomes). However, they can also
be used as delivery mechanisms, to put *new* DNA code inside the cell.
When used in this way they become *vectors*. The experiments here
all rely on making a vector that expresses BamA from a plasmid, in

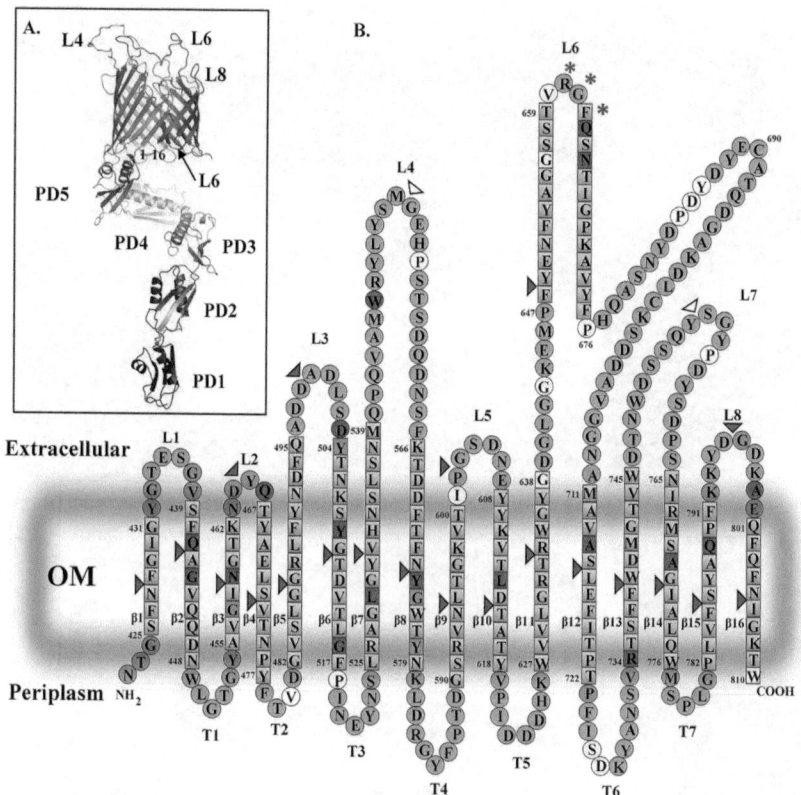

Source: Browning et al. (2013a: 6).

Figure 2.3 Topology model of the BamA β-barrel. Key features to note are the eight extracellular loops (L1–L8) and the sixteen strands in the OM (outer membrane).

contrast to the cell's usual way of making BamA from its main DNA. Plasmids are relatively straightforward to make and so important in molecular biology that it is often easier simply to choose one from a catalogue and either use it immediately or adjust it to your own specifications. In the case of this experiment the team used a ready-made plasmid as a vector – called pET-17b – and then modified it a number of times.

By inserting the vector, the team would make bacterial cells that expressed BamA from the plasmid, but would also express BamA from the main DNA. They needed a way to make BamA from the plasmid *only*, and they did this by growing the bacteria in a special medium. *E. coli* needs a sugar called arabinose to make BamA from its main DNA; by growing the bacteria in the absence of arabinose the team

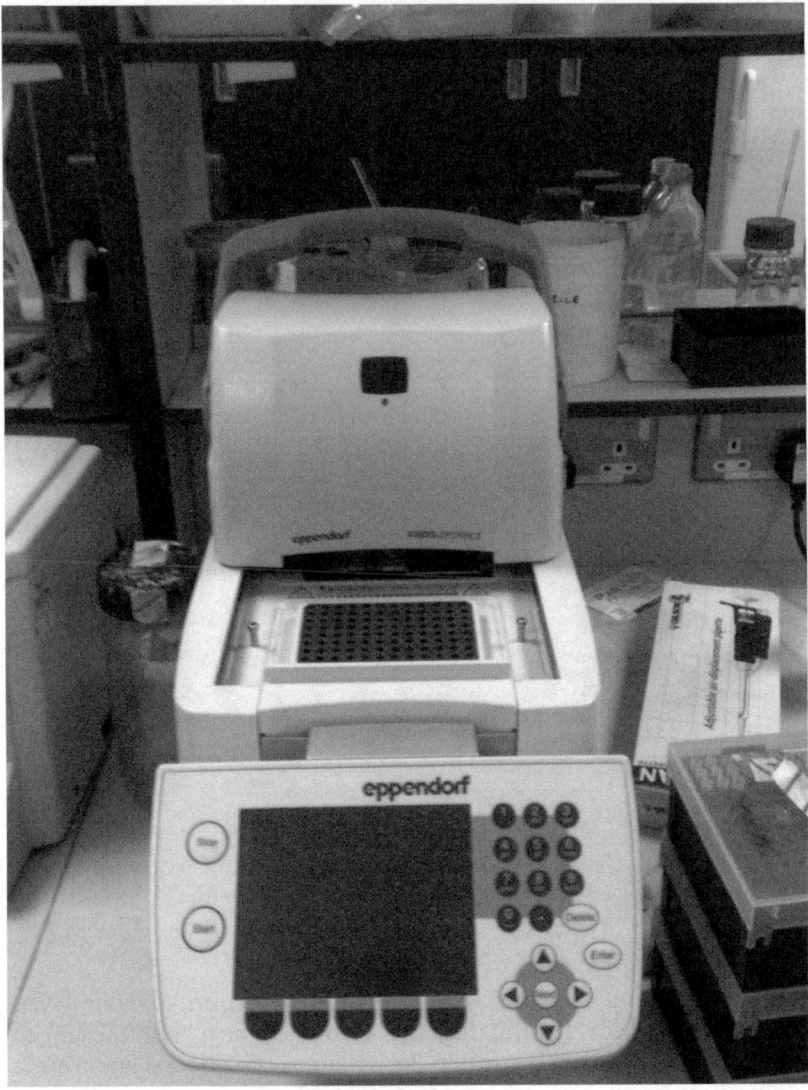

Figure 2.4 PCR machine

knew that all the cells were making BamA from their plasmid vector, not from the main DNA.

The aim of the team was to investigate the structure and function of BamA using mutagenesis, and they proceeded by constructing three linked experiments which together would provide inferences about the protein. The experiments required their own design, but were all carried out with reference to each other.

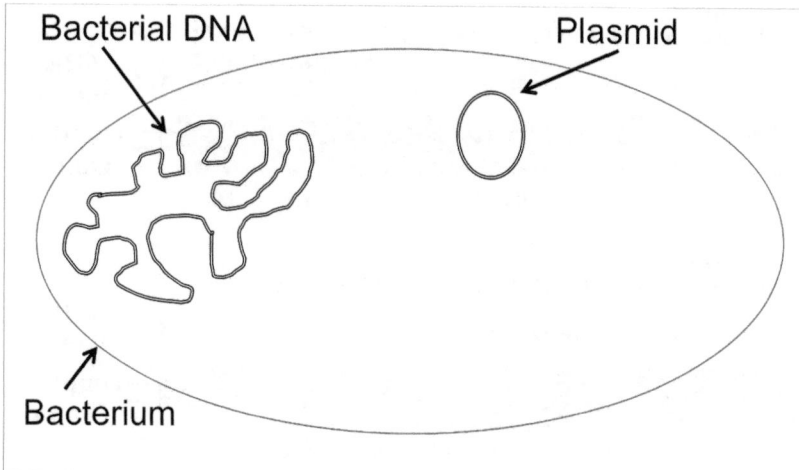

Figure 2.5 Schematic diagram of bacterium, DNA and plasmid

Experiment 1: linker scanning

The team knew that their plasmid vector had the original wild-type code for BamA on it. Using a method called linker scanning mutation, they made gaps in the plasmid at random and inserted a short additional sequence of DNA so that all new versions of BamA would have an extra insertion in them. These insertions meant that when the gene was expressed a short additional sequence of amino acids would be added into BamA at random points. The new versions of BamA could be compared for function and inferences drawn about the structure of the molecule. Each of these versions would be examined and the location of the insertions mapped (see figure 2.6). The team also measured how well the new versions of BamA worked. They did this by producing a *depletion strain* (explained in more detail later: see 'Phase 2: linker scanning mutagenesis of BamA').

Experiment 2: HA epitope insertion

As well as randomly altering BamA, the team wanted to look at specific regions of the BamA complex. To do this they used PCR to make new versions of the vector, specifically targeting each of the eight loops and sixteen strands of BamA (see figure 2.3). The new vector inserted an epitope – a little tag that would attract a paired antibody – called HA (human influenza hemagglutinin) into these specific regions. Again, the resulting cells were cloned and grown, then the depletion strains were examined to see how well they grew in the presence of the antibiotic vancomycin. By subsequently adding

Science, Culture and Society

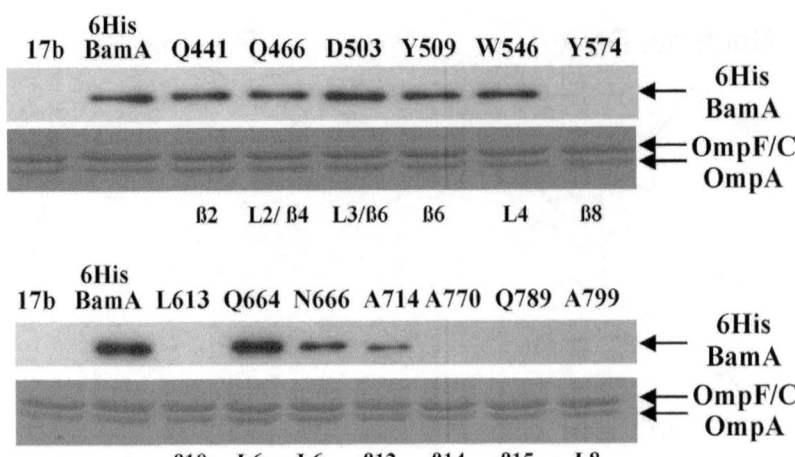

Source: Browning et al. (2013a: 7).

Figure 2.6 Mutational analysis of BamA β-barrel. This image shows dark bands of protein against the light background of the gel medium they are suspended in. Each band is a different mutated version of BamA.

a matching antibody that had been specially constructed to include a tag that fluoresced in the presence of ultraviolet light, the team were able to make images of the bacteria that showed which bits of BamA β-barrel were protruding through the outer membrane (see figure 2.7).

Experiment 3: deletion of BamA surface loops

Once they knew which loops of the BamA complex stuck out of the outer membrane, the team could begin to determine what the loops were for, and how important they were to the functioning of the whole complex. The team did this by making new plasmids that coded for the whole of BamA except for, sequentially, individual loops (see figure 2.3). Once again, the new cells were grown in the absence of arabinose, and then in the presence of vancomycin.

So there were three separate, but linked, experiments, each of which would allow inferences to be drawn about the structure and function of BamA. But the brief descriptions provided, which are translated out of the Browning et al. (2013a) paper, hide a great many things. We need to look in a bit more detail at the actual practices in the laboratory that are making this knowledge. How does experimental design become experimental practice?

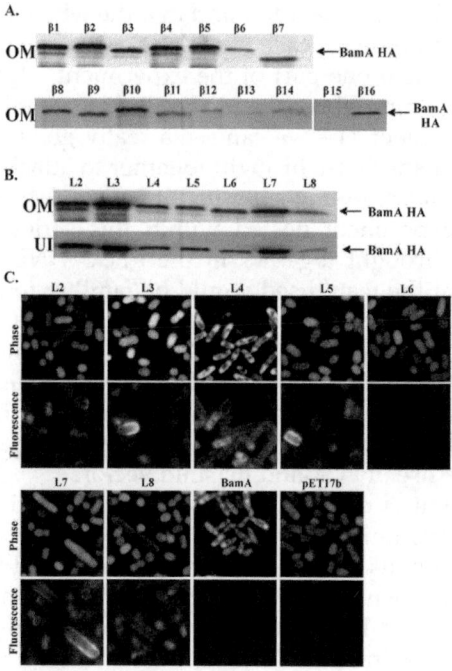

Source: Browning et al. (2013a).

Figure 2.7 Analysis of HA epitopes within the BamA β-barrel

From design to practice to results

Looking at the *PLOS ONE* paper's 'Materials and Methods' section reveals a fundamental issue in formal science: the difference between the written account of the experiment and the actual physical and mental labour that has gone into doing the experiment. In the paper, although all the necessary technical details of how the experiments were done are provided such that another team of experimenters could reproduce the experiment exactly, there is no indication of the process by which the work was carried out. Putting this crudely, the human factor of thought and effort is entirely missing from the account given in the formal science paper. Where, for example, the paper describes 'plasmid construction' in a couple of paragraphs, and all the technical details are there, we get none of the process of very complex, meticulous and patient laboratory bench work that is required to produce so many different constructs. The sheer volume of this work is significant and it is worth just considering stepwise what this method requires in detail to provide some idea of how complex, lengthy and meticulous this part of the experiment is. Bear in mind that the whole paper is bringing together three experiments, each of which needs similar human effort.

If we focus on just one part of the experiment – 'generation of the BamA linker scanning library' and the 'linker scanning mutagenesis of BamA' (experiment 1) – we can see a really good example of how these different methods are brought together to attack the problem. In addition, the linker scanning experiment is complete in its own right, a stand-alone experiment nested within this series of experiments that have been brought together in the *PLOS ONE* paper. Some of the methods that the team used would be familiar to a microbiologist working in the 1930s: growing bacteria on solid media (agar gel in Petri dishes) or in liquid media (broth). Alexander Fleming used similar methods in his laboratory when he discovered the antibiotic properties of penicillin. Molecular microbiology still uses these techniques, but augments them considerably by applying techniques from genetic engineering: the use of PCR and plasmid vectors.

The experiment is designed to insert 5 amino acids (the building blocks of proteins) randomly into the BamA protein so that the team can see what happens to BamA when it is disrupted in this way. We can see this experiment as consisting of two phases: making the linker scanning library, and linker scanning mutagenesis of BamA. We will look at each of these in turn.

Phase 1: making the linker scanning library

The team knew that their plasmid vector pET17b/*bamA* had the original wild-type code for BamA on it. Using linker scanning mutation they made gaps in the plasmid at random and inserted a short additional sequence of DNA – fifteen base pairs (15bp) – so that all new versions of BamA would have an extra insertion somewhere in them.

The team used a tried and tested technique that has been turned into an 'off the shelf' kit (the Thermo Scientific Mutation Generation System (MGS) Kit – 2013 cost $170 each, and two kits were needed for these experiments) to facilitate the process. The kit is a standard genetic engineering tool in that it relies on basic recombinant DNA (genetic engineering) principles discovered in the 1970s from work on viruses. Inserting a piece of DNA into an existing strand or loop of DNA makes a 'recombinant DNA molecule', and this is how viruses work: they infect a host cell and integrate their DNA into the host's DNA. Biochemists recreated this process in vitro in the 1970s. In essence, plasmids are isolated, then, using specific enzymes, they are cut open, additional pieces of DNA are inserted and, again using enzymes, the circles are closed up. Linker scanning works by taking the target DNA, cutting it open, inserting an entranceposon (an artificial sequence of DNA that can change its position in the main DNA sequence; a linker), then expressing this in *E. coli* cells by growing them on. From this collection of new *E. coli* cells, the team could extract the DNA and determine if their target piece of DNA is on the plasmid or the main gene. Using an enzyme (NotI5) the team cut out the entranceposon and then resealed this gap. The plasmid now had the 15bp sequence that would code for the five amino acid insertions. These 15bp insertions meant that when the gene was expressed, a short additional sequence of amino acids would be added into BamA at random points. Again, this new DNA was inserted into *E. coli* cells and the bacteria grown on. Finally, the team extracted the DNA from these new cells and sequenced it, thus generating the library of mutations they had made using this linker scanning technique. However, the previous couple of sentences, like the *PLOS ONE* paper, hide an incredible amount of detail.

Most laboratory workers would agree that what their work involves is following complicated recipes in precise ways. This protocol was no different; it required a high degree of skill, precision and patience, good organizational ability, lots of planning at the outset, and a certain amount of luck. Here's the actual stepwise process by which this part of the experiment described was done. I'm including this to show the sheer volume of precise work that had to be carried out to produce the results:

- The experiment was based on a series of chemical reactions – enzymes cutting DNA or sticking DNA back together using known chemical reactions. These reactions had to be confined and controlled as tightly as possible. The reactions in this experiment all took place in microcentrifuge tubes, more widely known as Eppendorf tubes (see figure 2.8).
- Step 1: put the target DNA (the team's vector pET17b/*bamA*) into the tube with the enzymes and reagents supplied in the kit using a micropipette. Mix and incubate, then deactivate (see figure 2.9). This will have made the transposition complex.
- Step 2: insert this DNA into *E. coli* cells using electroporation (an electrical pulse blasts the cells open and the plasmid enters).
- Step 3: plate the bacteria out onto agar plates and grow them on (see figure 2.10).
- Step 4: collect single colonies of the bacteria (each agar plate will have lots of colonies, but each will have a different version of BamA in it). Grow the cells overnight in liquid medium so there are enough cells to isolate sufficient DNA.
- Step 5: use an off-the-shelf Qiagen miniprep kit (QIAprepSpin Miniprep kit (2013 cost, £63.50)) to break open the cells' membranes and extract and then purify the DNA using a small spin column.

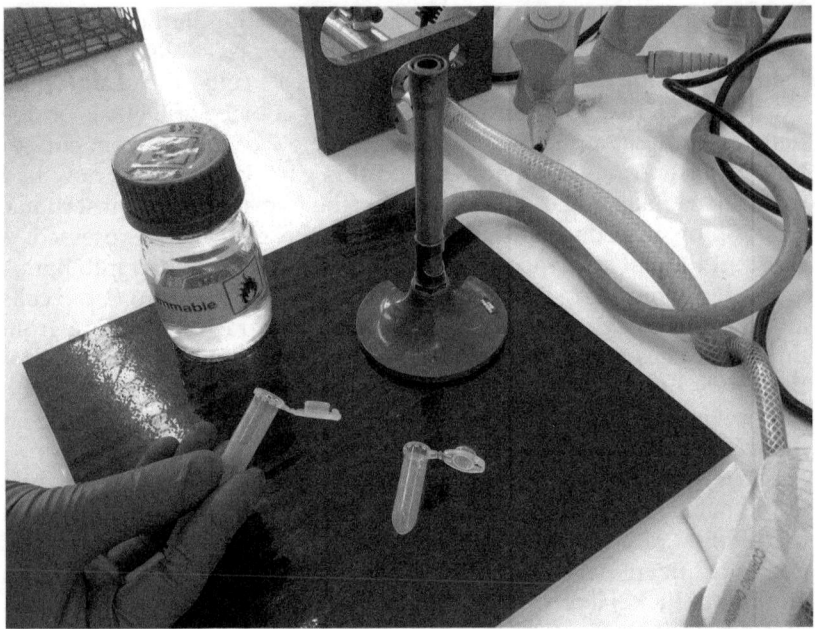

Figure 2.8 Eppendorf tubes

- Step 6: remove the Entranceposon insertion using a standard restriction enzyme (NotI).
- Step 7: use gel electrophoresis, to separate and isolate the different bands of DNA by applying a mild electric current which 'pulls' the lighter-weight DNA molecules across the gel faster than the heavier ones (see figure 2.11).
- Cut these out of the gel using a scalpel.
- Step 8: using remainder DNA extracted at step 5, turn this back into plasmid (circle) form with an enzyme, and insert into competent cells, as in step 2 (figure 2.12).
- Step 9: grow these cells in a depletion strain by including the anti-biotic vancomycin in the agar medium.
- Step 10: collect results by looking at which bacteria grow, and how well they grow.

Those ten steps have been greatly simplified for the purpose of clarity and concision: the reaction protocol for the Thermo Scientific Mutation Generation System (MGS) Kit is available online (http://www.thermoscientificbio.com/uploadedFiles/Resources/tech-manual-f-701-mutation-generation-system-kit.pdf), and runs to twenty-four pages. But this protocol is described in the *PLOS ONE* paper in this way:

Figure 2.9 Using the centrifuge

The BamA linker scanning library was generated using the Thermo Scientific Mutation Generation System Kit. Entranceposon M1-Kan^R was randomly introduced into plasmid pET17b/*bamA*, as specified by the manufacturers, and transformed into *E. coli* K-12 strain RLG221. Insertions were selected for by plating cells onto LB agar containing 50 µg ml^{-1} kanamycin and then screened for insertions within *bamA*. The entranceposon was removed by digesting each plasmid with *Not*I. Restricted plasmids were re-circularised using T4 DNA ligase and transformed into cells selecting for ampicillin resistance. The location of each 15 bp insertion within *bamA* was identified by DNA sequencing and all insertions are listed in Table S3. (Browning et al. 2013a: 2)

These ten steps are reduced to six sentences (106 words, to be precise). Note as well that in all the team made *eighty-seven* new versions of BamA

Figure 2.10 Plating out bacteria

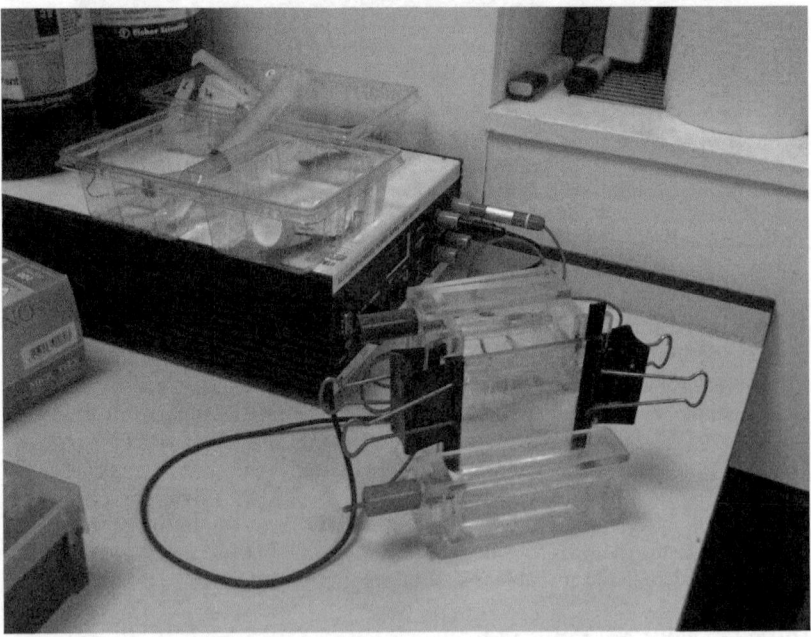

Figure 2.11 Gel electrophoresis apparatus

Figure 2.12 Breaking cells open using the electroporator

which they could compare for function in order to draw inferences about the structure of the molecule. Each of these versions was examined and the location of the insertions mapped using DNA sequencing.

As with the *PLOS ONE* paper's description of plasmid construction, the description of generating the BamA linker scanning library is technically precise, but disingenuous in terms of the sheer volume of work that was needed to do this. After all, eighty-seven different bacterial strains were generated in this process, each one of which needed to be isolated, plated out onto solid media, grown, selected and put into liquid media, grown, assayed, then isolated again, grown on solid media again, this time in the presence of an antibiotic, and then assayed. Then each experiment had to be repeated at least once – twice or three times is more common – to check that the results were correct. Dr Browning estimates that the linker scanning work for the *PLOS ONE* paper took approximately nine months of work for him and three other team members to carry out. In the paper, that huge amount of work is reduced into those 106 words.

My ten-step account considers the actual work and physical practices; Browning et al.'s account focuses on the essential parameters of the experimental conditions that distinguish their experiment from any other experiment deploying the MGS kit; the Thermo Scientific Mutation Generation System (MGS) Kit protocol focuses on essential information to make the kit work. All three accounts are of the same thing, but the focus and emphasis are very different in each one as the accounts are oriented to specific audiences with specific shared thought.

Phase 2: linker scanning mutagenesis of BamA

The team also measured how well the new versions of BamA worked. They did this by producing a depletion strain, a technique that compares the functional ability of different cells that have been depleted in one way or another through genetic mutation. They grew each of the eighty-seven different versions of the bacteria in a medium that did not contain arabinose, so that they knew that the bacteria were making BamA from the plasmid only (see figure 2.13).

Not all of these cells could grow, as the version of BamA they were making was not functional, so the cells died. But those that did grow, the viable cells, were subsequently grown on agar plates that contained the antibiotic vancomycin. Vancomycin is quite a large antibiotic and it needs a big 'doorway' into a cell for it to kill the bacteria; *E. coli* K-12 is normally insensitive to high concentrations of vancomycin. By growing the bacteria in the presence of this antibiotic the team could see if the cells had a version of BamA that functioned well, as it would make an efficient cell wall, with properly folded and working outer membrane

Figure 2.13 Preparing the depletion strain

proteins, for *E. coli* and this would keep the vancomycin out; if the BamA was not so well-formed and functioning then vancomycin could get into the cell and either retard growth or kill the bacteria.

What did the experiments show?

The results tell us a lot about BamA, and by extension about Gram-negative bacteria, the class of bacteria that *E. coli* belongs to, as a whole. The team split the results into two sections: those relating to the C-terminal of the protein which is embedded in the cell's outer membrane, and those relating to the N-terminal which is inside the bacterium. A third set of inferences, about the loops that stick out of the cell, is also included.

At the C-terminal the experiments showed that making insertions into the protein 'destabilizes the BamA β-barrel and is consistent with our topology model' (Browning et al. 2013a: 4). In other words, disrupting the wild-type sequence of amino acids that form the β-barrel in the outer membrane makes a big difference to how the protein works. But the fact that *some* insertions could be expressed allows the team to draw inferences about how *E. coli* makes BamA.

At the N-terminal there are five polypeptide transport-associated motifs ($POTRA_1$ to $POTRA_5$). Previous experiments into these parts of BamA had suggested that only $POTRA_3$, $POTRA_4$ and $POTRA_5$ are essential. However, Browning et al.'s results found that 'all five POTRA domains are essential for normal laboratory growth and viability' (Browning et al. 2013a: 1).

By making insertions into the DNA code for the six loops of BamA that protrude from the outer membrane the team could determine by inference which loops are necessary. In this case they found that insertion tests suggested that loops L2, L3, L4 and L6 are likely to be important.

The experiment successfully addressed its aims and provided a huge amount of information about this complex protein. But it did not resolve all questions about BamA, and the paper concludes that further mutational studies could 'reveal the intricacies of this complex molecular machine and unravel the mechanism by which the BAM complex folds and inserts OMPs [outer membrane proteins] into the outer membrane' (Browning et al. 2013a: 10–11).

Formal science work, language and discourse

The description I've given of the experiment is a 'translation' of two more formal scientific discourses. The first is the discourse that

I collected in the laboratory, listening to (and watching) one of the scientists (lead author Dr Doug Browning) carrying out his share of the experiments. The scientists in the team discussed the experiments and other projects they were working on, and also discussed what the results of experiments meant. In other words, they shared their inferences drawn from the results and their construction of the context that the results should be understood within. Such discussions take place over a long period of time, often years, as the work is complicated and experimental design requires a lot of thought. In the case of the experiments under discussion here the whole time scale, from writing the grant application to publishing the paper, was about six years. Laboratory work is a bit like an iceberg: the visible portion is what can be seen in the publication or conference presentation, but beneath that tip there is a massive amount of preparatory work, failed experiments, grant writing, meetings, ordering lab supplies, writing draft papers, emailing colleagues in different parts of this thought community and so on.

Box 2.4 Formal scientific language

The Browning et al. (2013a) paper collects together a number of experiments and presents a range of results. The following quotation illustrates the format that discussion of complex experimental results will often take:

> Our linker scanning mutagenesis isolated a number of insertions in both the POTRA and β-barrel domains of BamA, which led to severe phenotypic defects and indicate that both domains play a role in OMP biogenesis. Interestingly, all severe insertions within the POTRA domains were associated with $POTRA_2$ and $POTRA_3$, whilst none were within $POTRA_1$, $POTRA_4$ and $POTRA_5$, even though these POTRA domains are essential (Fig. 1) [6]. Due to the methodology of library construction (see Materials and Methods) it is unlikely that the failure to isolate such mutations was due to toxicity but rather reflect that our BamA mutagenesis had not reached saturation. Our observation that all POTRA domains are essential is in contrast to those of Kim et al. [6], who initially showed that $POTRA_1$ and $POTRA_2$ were dispensable for BamA function, although cells bearing these deletions grew extremely poorly. (Browning et al. 2013a: 7–8)

Summarizing complex experimental results is a difficult task, but Browning et al. do this in a few sentences and place their results in a wider context by making explicit reference to other recent work on BamA's POTRA domains that their work contradicts (i.e. the work of Kim et al. 2007, cited as source [6] in the extract above).

The second scientific discourse we can identify in the paper articulates the project of science. The paper starts with an identification of what is already known about BamA and attendant systems through citing the relevant literature. It is notable that the paper doesn't cite any of the 'paradigmatic' literature: Watson and Crick's work on DNA, the invention and application of PCR and so on. These simply do not need to be mentioned: they are the tacit knowledge (Polanyi 1958; Collins 2010) of this thought community. The paper proceeds to show what has been done and what has been found, then points in directions of future work in this area. The small end point that the experiments aimed for – addressing the research questions – has been achieved. But the scientific discourse expressed in the paper is showing that another end point – the complete knowledge of the natural world – is achievable and will also be achieved. This is the core tenet of the scientism that is abroad in contemporary society.

Some observations on these experiments

The biosciences laboratories where this group of experiments took place are busy working environments where teams of researchers will spend weeks at a time working at lab benches carrying out experiments, often repeating the same processes again and again. But the lab is more than just a place where experiments are repeated. There is a constant stream of conversation, discussion and argumentation between researchers, with individuals sharing thoughts, ideas and suggestions for how to do their work. This is a fertile and creative environment, a place where a thought community does a lot of thinking and talking, and the informality of the interactions is probably a contributory factor in this. The level of activity and the high work rate in the laboratory mean that work must be closely co-ordinated – for example, workers need to book the 'hot' room to work with radioisotopes – and the ability to co-operate and work as part of a team is essential. Despite this there is a certain amount of disorder in the laboratory, and in two ways. The first is the disorder of social interaction: laboratory workers interact and discuss – often quite heatedly – their work and the work of others. Discussions and arguments about what people are doing, why they are doing it and what they should do next are vital for the production of this formal scientific knowledge. Whilst the main ideas and thought style of this esoteric community are shared and don't need to be discussed, individual workers' experiences are varied and are the basis upon which experimenters will be designing and executing their experiments. The second kind of disorder comes from the intrusion of the outside world into the esoteric laboratory environment: lab benches become decorated with personal photos and mementoes; incongruous

Figure 2.14 Ice cream tub in the 'hot room' (radioactive isotope work area)

objects enter the laboratory (see figure 2.14) and spaces become disordered as rubbish builds up (see figure 2.15).

Although the experiment is complex and requires a great deal of precision, it can be completed with fairly basic equipment and does not require a great deal of space. The main equipment required is a PCR machine – which, although an almost miraculous piece of kit, is

Figure 2.15 Rubbish building up in the laboratory

now absolutely standard in any biosciences lab and a basic unit can be bought for about £1,000 – agar plates and liquid media to grow bacteria on, and basic lab tools such as pipettes, Bunsen burners, test tubes, latex gloves. Much of the equipment in the laboratory – refrigerators, microwave ovens, sinks, water purifiers – would not be out of place in an ordinary kitchen. The experiments themselves do not take much time: a day of preparation, a day of running the experiment and a day of looking at results. But each experiment must be designed and refined through testing and this can take weeks or months. Once the experimental protocol has been decided the experiment must be repeated at least twice to check the results are correct. In the case of the BamA experiments this meant a huge amount of repetition, and the experimenters had to be very well organized to keep track of a large amount of data coming from each run of the experiment.

These experiments were funded through a number of different sources. All of the team were employed by the University of Birmingham, although the income streams to pay them come from different places. Two research councils (government bodies that distribute money to, mainly, universities for academic research purposes), the Medical Research Council and the Biosciences and Biotechnology Research Council directly funded the research. UK funding councils administer very large budgets and distribute it in two main ways. The first is by giving money to teams of academics who bid for their projects to be funded. This bidding process is highly competitive and is subject to lengthy reviewing processes. The second is by directly funding specific institutions and facilities. The UK research councils are funded directly from general taxation. The main three science funding councils' budgets in 2012–13 were: BBSRC (Biotechnology and Biological Sciences Research Council) £467m, EPSRC (Engineering and Physical Sciences Research Council) £866m, MRC (Medical Research Council) £766m. However, the Ministry of Defence (MOD) is the 'highest spending UK government department on R&D activities (representing around a third of all net R&D funded by UK Government), and internationally ranks second only to the United States in the level of R&D investment undertaken on defence tasks' (Stone and Bennett 2009: 1). The MOD's net expenditure on R&D in 2011–12 was £1.3bn, of which 90 per cent was spent extramurally, i.e. in HE institutions, industry or overseas (Ministry of Defence 2013). The research described in this chapter was not funded by the MOD.

The laboratory is organized into teams inside which operates a hierarchy. Team leaders are generally lecturers or professors, i.e. they are permanent academic staff whose duties include teaching and carrying out research. The team members are, generally, researchers only, although they may have a small amount of teaching duty including supervising undergraduate and postgraduate students in laboratory sessions. It is this group of staff – postdoctoral researchers (colloquially 'postdocs') – who are the main producers of formal scientific knowledge and scientific papers in the UK. Finally, the team may be augmented by PhD students, who are attached to specific projects and who are primarily carrying out experiments as part of their PhD training and for use in their PhD thesis. Postdoctoral researchers are arguably the most highly qualified group of staff in the UK labour force, excluding medics, having at least an undergraduate honours degree and a PhD (i.e. a *minimum* of six years HE), and are responsible for producing the vast majority of scientific outputs. Despite this their employment conditions are often poor. The vast majority are employed on temporary contracts, typically lasting the length of the research council funded project they are attached to (Erickson 2002), which has significant effects on their chances of career progression. Postdoctoral researchers are less likely to be members of a trade union than university teaching staff, and have much less representation in university structures and committees.

Outcomes

What happens to the results from these experiments? The first, obvious outcome is the scientific paper that was published in December 2013 in *PLOS ONE*, a peer-reviewed open-access journal (see box 2.3 on p. 36). The scientific paper is an incredibly deracinated object that has deliberately stripped away almost all vestiges of the human side of the process that has created this scientific knowledge. There are reasons for this, the most obvious being that it needs to conform to the genre it is a part of. Scientific papers are written in this way because, first and foremost, they are just written in this way and the scientific community is conservative, averse to change and particularly averse to including anything that makes it look as if the process by which scientific knowledge emerges is anything but wholly objective. The published paper doesn't describe everything that takes place in the production of the results of the experiment. It is an edited and truncated description of the experimental procedure which focuses on the successful experiments, not the unsuccessful ones, and which only describes the absolutely necessary information for someone skilled in a similar range of techniques on a similar range of objects to make sense of the results. Significantly, the paper omits general discussion of the principles of polymerase chain reaction, the structure and function of DNA and RNA, the general scheme of outer membranes in bacteria: it is assumed that the audience of the paper will know all of this information already. Effectively the paper is a translation, and a condensation, of a large amount of formal and informal interaction (e.g. discussion between researchers in the laboratory) that has taken place over a number of years. The point here is that it is the laboratory practice of working for an extended period in a highly specialised area, designing a large number of experiments (many of which will be failures), refining techniques through discussion with colleagues in a team and in other laboratories, confirming and rejecting hypotheses, and moving from existing theories on a subject to extend knowledge of an area that is the formal scientific method. The scientific paper is a pale reflection of this process, and a reflection that hides key features of the collective and non-linear processes that make up scientific research work.

The paper will inform other researchers working on similar experiments that are investigating β-barrel structures in bacteria and in eukaryotic cells. Other researchers may choose to replicate some aspects of Browning et al.'s work to confirm the results (this is what

Browning et al. themselves did with Kim et al.'s 2007 work), although it is unlikely that the exact experiment will be repeated. It is more likely that Browning et al.'s work will provide a new starting point for other researchers.

As well as providing new information about BamA, the experiments have also confirmed that the techniques and procedures used are good and appropriate. This is in some ways as important as the information about BamA: researchers could choose a range of methods to look at BamA, and there are many different types of BamA that need to be investigated. By choosing and confirming the worth of their method the researchers are increasing the confidence of their esoteric thought community that this is one of the best ways to investigate these outer membrane proteins.

A third outcome is confirmation of the models being hypothesized. At the most basic level, the results of this experiment confirm that DNA is transcribed to make RNA and this is translated into proteins. The experiment directly links to the overarching paradigm of the biological sciences and confirms that the paradigm appears to be correct (see chapter 3).

Formal science

From the outside, formal science can look like a number of very large projects that are working on big 'problems' – the problem of gene therapy, the problem of what matter is really made of, the problem of bacterial resistance to antibiotics, the war on cancer and so on – but from the inside it becomes clear that there is in almost all cases a very large 'gap' between an individual experimenter's work and the big 'problem' that their work is trying to address. Science in our present society proceeds by breaking down big problems into ever smaller problems, solving these as best as possible, and then amalgamating (if possible) results from a whole series of small-scale experiments to try and shed light on the original big problem. However, even this abstract description makes it appear as if science is a kind of hierarchical conspiracy, where there are some people who are controlling the research strategy as a whole. This simply is not the case: bringing together the elementary researches of the mass of scientists involved in a broadly similar field does not take place in a systematic way. Rather, it takes place in a fairly haphazard way (particularly if we take a global view of formal scientific endeavour), with new research projects combining previous work with new experiments in the hope that a synthesis of ideas will produce useful results. The complexity of what formal science as a whole is trying to do can be illustrated by considering an example.

If we take a disciplinary area like biochemistry, and look at a significant problem that biochemistry is trying to tackle, such as the emergence of bacterial resistance to antibiotics with a view to providing a new generation of more effective antibiotics, we will find that a huge number of different projects are involved in this endeavour. Some projects will appear to be 'closer' to the main topic than others simply by virtue of their self-descriptions, and these may or may not be the projects that will bring about the next generation of antibiotics. All knowledge, all information relating to how bacteria develop resistance to antibiotics and what sorts of chemicals inhibit bacterial growth, will clearly be of use in tackling and solving the big problem. However, it may be that the next generation of antibiotics emerges from work being done in molecular microbiology, by investigating how bacteria replicate themselves. New antibiotics may not look anything like old antibiotics (indeed, it is even possible that the next leap forward in antibacterial treatments could come from nanotechnology, a branch of physics, as argued by Eric Drexler (Drexler 1990)). The complexity of trying to resolve one, on the surface, fairly straightforward problem is revealed only when we really start to break down the problem. The current focus on the biochemical approach to producing new antibiotics is a reflection of the sense in science and society that genetic knowledge is the most appropriate knowledge for changing, and improving, our world. (See Lewontin 1993 and Nelkin and Lindee 1995 for a more in-depth analysis of these themes.)

Looking at one specific research project in particular will reveal very little about the problem as a whole, but will tell us a lot about why a complete understanding of this problem has not yet been found, and why it is that science is so hard to make sense of.

It appears that the natural world is fairly straightforward to understand because we are told that it works according to rules and laws, and we think that the social world is much more difficult to make sense of because we cannot identify those rules and laws. What we fail to notice is just how incredibly complicated the natural world actually is: simply having the rules and laws helps very little when we are dealing with complex and disparate phenomena. Some sociologists of science have described this everyday laboratory work as 'puzzle-solving' (see section on T.S. Kuhn in chapter 4), but really what we are seeing is the application of proven tools to new areas of inquiry to produce descriptions of the workings of, for example, the bacterial cell. There is no puzzle here in the sense of there being a theoretical prediction that can be tested and, if it doesn't match up to expected results, will be, perhaps eventually, falsified. What most scientists have is a set of laws, rules, techniques and prior examples that can be applied to the vast number of cases that have yet to be investigated, extending their current knowledge and expertise into new examples. Knowing that the production of one protein is regulated by another enzyme does not tell us about how a different protein is regulated, not to mention how two proteins are regulated separately when they are being produced simultaneously inside a cell, although it may provide us with clues that will help us investigate further cases.

Different kinds of formal science

Although our common understandings of science and our experience of formal science at school tend to focus on experiments, not all scientific endeavour proceeds by the method of testing hypotheses through controlling variables and finding experimental results. Experimental sciences do this. However, there are some scientific activities that simply can't proceed by testing things in experiments, yet they are also classified as being 'science' by most observers: we generally call these discovering sciences. An example is astronomy, and other 'discovering' sciences are geology, botany, palaeontology. However, all of these will use, at times, experimental technique, and a separation between experimental and discovering sciences is in many ways an artificial one. Apart from anything else, results from observations are essential to constructing new theories which generate ideas for new experiments. We should also note that many commentators, indeed many practising natural scientists, would consider the social sciences of sociology, psychology, economics and politics to be 'discovering' sciences. Much of what astronomers do concerns observing things, cataloguing these phenomena, and then drawing conclusions that try and provide answers to the 'why' questions that are raised by the phenomena. Astronomers are 'discovering' things about how the universe works through observing it, albeit in highly systematic ways. For example, astronomers may observe some new phenomena in the night sky. They will then attempt to find similar examples, and will then try and find commonalities in terms of the data they have collected which will help to explain why the phenomena have the characteristics that they do. We can note that significant characteristics of this form of method would be the attempt to carry out observation systematically, to measure phenomena as precisely as possible, to record data as accurately as possible, and to apply existing rules to results, or to construct new rules that can explain the main features of the phenomena in question; all aspects of a general scientific method.

The discovering/experimental dichotomy is by no means the only distinction that we can use to categorize scientific enterprises. Most people will be familiar with the distinction of scientific sub-disciplines into the field of life and physical sciences. These distinguishing labels are visible not only in media descriptions of the sciences, but also in university structures where biology may be located in a faculty of life sciences with disciplines such as psychology and pharmacy, as distinct from faculties of physical sciences containing disciplines such as chemistry, physics, astronomy. The split between life and physical sciences is becoming harder to maintain, or even identify, in contemporary formal science. Biology is increasingly being dominated by biochemistry – the chemical study of life processes that often proceeds by experiment – and disciplines such as chemistry and physics are increasingly concerned with the construction and operation of biological systems.

A further twist to these distinctions is provided by scientific endeavours that neither carry out experiments nor observe the natural world. Formal science that proceeds by creating computer-generated scenarios and analysing the outcomes of changing variables in these are not particularly new, but are much more common now. It is quite possible for scientists to present research findings on new chemical compounds that have been 'created' entirely inside the memories of computers. As noted previously, this presents a challenge to those who consider 'science' to be about 'nature'.

What are experiments for?

The experiment we looked at shows that laboratory practices generate specific data about specific phenomena. The data collected in the experiments is used to infer knowledge about the objects in question. In this case, the results of the molecular biology experiments are used to make inferences about the structure of BamA. We can therefore clearly see that one of the stated goals of the project of science – the expansion of certified knowledge of the natural world – would appear to have been achieved by this experiment. However, some philosophers of science would question the validity of the knowledge being produced, in that they would challenge the reality of the objects being studied. This is not something that most scientists would do, and in the case of the experiment looked at here the experimenters consider that the objects of inquiry – BamA – actually exists and can be considered to be 'real' because 'real' data is produced about it. For the moment we will suspend judgement on this point. The connection between the data collected in experiments and the knowledge that is produced in scientific publications and other forms of scientific discourses will be discussed in the next chapter.

Although it can be said that a goal of the project of science has been achieved by producing and extending certified knowledge of the natural world, this may not be sufficient reason for carrying out the experiments. These experiments result in publications that will inform a small part of the scientific community, but the knowledge that appears in the publications will, in all likelihood, not be used by anyone immediately, and may not be used by anyone ever, or at least not for a considerable period of time. This does raise the question of why anyone should bother to carry out the experiment in the first place, and considering this question in the light of understanding what the experiments entail shows something important about contemporary science.

The molecular biology experiment is at the cutting edge of the discipline: it provides new knowledge of the world that has not been seen by people before. The researchers carrying out these experiments are part of quite small networks of researchers around the world, an esoteric thought community, who are carrying out similar research using similar experimental techniques, and there are many such communities. These esoteric thought communities are engaged in complex experiments into very specific things that do not, at least immediately, have connections to other fields of study in their discipline. Unlike the representations of science that are often presented within popular culture, where experimenters are directly working on projects with specific applicability (e.g. a 'cure for cancer'), most science – particularly the science carried out in academic environments – is like the experiment shown here. Our societal representation of science is often to compare it to an army, advancing on a front in the, for example, 'war on antibiotic resistance'. A better metaphor is to consider formal science to be like a meadow and each esoteric thought community as a blade of grass sticking up through the soil. It may be that there will be specific applications that emerge from these experiments: the molecular biology experiments may lead to new forms of antibiotics being produced, but they may not. The breakthrough may come from a completely different part of formal science, or from a combination of different parts of formal science, or may never come at all.

What experiments are doing is extending the reach of researchers into new areas of study. In addition, experiments are also validating a particular experimental procedure, and validating a way of imagining the natural world. A further point that must be noted is that laboratories are training grounds for a new generation of academic scientists (see chapter 5), and a training ground for highly skilled researchers, computer programmers and managers who will go on to work in industry, the banking and finance sector or elsewhere (Lanchester 2010: 40; Bouchaud and Potters 2003).

This discussion has begun to address the issue of why these experiments are happening, but there is a final point that needs to be considered. These experiments are happening because there is a tradition of doing experiments such as this in these laboratories, and there has been a large amount of money, time and effort invested in previous experiments that the current experiments build upon. This inertia is hard to overcome, and there is little desire inside esoteric thought communities that are continuing on their research track to turn aside and pursue different avenues of inquiry.

Analysis of the work carried out in laboratories reveals the relationship between the results created in experiments and the narratives that emerge in scientific papers. In addition to the formal outcomes of scientific papers, the research process in the laboratory incorporates a range of social factors implicated in how experiments take place (e.g. team work, group discussion, informal chatting) and some external structural factors (e.g. funding regimes and the peer-review process) that affect what is done in the laboratory. A process of translation occurs in the transition between laboratory practice and description of formal scientific knowledge in a scientific paper. A subsequent process of translation is necessary on the part of the reader of that paper, aligning the written words with the experience and knowledge that the reader has. Overall, knowledge is being produced at a number of levels, and each with a different ascribed status.

Looking inside the experimental processes taking place inside laboratories reveals many things. Purely in terms of the descriptions used by researchers it is possible to note some distinctions. Initial consideration of topic areas and discussion with colleagues in laboratories are often quite informal, with loose ideas for methods and hypotheses being discussed in informal settings. This is translated into formal scientific method in the form of protocols for experiments. In the course of carrying out experiments, less formal practical features emerge, as researchers adopt pragmatic 'fixes' to get equipment working and ensure results emerge. Results are formalized by analysis, and experiments are subsequently presented in the highly formal setting, and highly formal language, of the scientific paper. Introducing a social researcher into this flow entails researchers having to translate their internal, esoteric descriptions – the dialect of their thought community – into exoteric general language that outsiders can understand: again, this is a shift between formal and less formal modes of explanation.

The next chapter will begin to examine the next aspect of doing science: how formal scientific knowledge is understood by commentators, what the status of formal scientific knowledge is, and what the relationship between formal scientific knowledge and wider understandings of science is. This involves considering the epistemological and historical aspects of formal science, and the construction of general histories and epistemologies of science and scientific knowledge.

Further reading

For insiders' accounts of scientific work and method a good starting point is to look at general introductory-level textbooks for undergraduates. Your local or university library will have a good range of these (look for one-word titles such as 'Biochemistry' or 'Physics' and try to find up-to-date editions). Almost all of these will contain brief descriptions of what scientific work and method are. It is interesting to compare these descriptions to the versions written by social investigators of science, such as those provided in the next chapter of this book.

The best textbooks for understanding the cell science and genetic analysis in this chapter are:

Becker, W.M. (2009) *The world of the cell*, London: Pearson/Benjamin Cummings.

Griffiths, A.J.F., Wessler, S.R., Lewontin, R.C. and Carroll, S.B. (2008) *Introduction to genetic analysis*, 9th edn, New York: W.H. Freeman.

Horton, H.R., Moran, L.A., Scrimgeour, G., Perry, M. and Rawn, D. (2006) *Principles of biochemistry*, Upper Saddle River: Prentice Hall.

Nature and *Science*, both weekly international journals, publish scientific papers from a wide range of disciplines. Almost all of these will include extensive description of methods used in the experiments or observations that provide the basis for the scientific paper. Both journals also contain 'science' news and op-ed features. *Nature* was founded in 1869, *Science* in 1880. Much more recently, and part of the open-access shift in science publishing, is *PLOS ONE* (PLOS stands for 'Public Library of Science'), started in 2006: this online-only journal (www.plosone.org) publishes peer-reviewed papers from a wider range of scientific disciplines. Unlike *Nature* and *Science* (and most other journals) *PLOS ONE* does not use the importance of the material as a criterion for publication, but decides only according to the quality of the research.

Antibiotic resistance

You will find frequent articles on this topic in journals such as *Nature*, *New Scientist*, *Scientific American* and *Science*. For a comprehensive overview of the topic the CDC 2013 report is excellent:

Centers for Disease Control and Prevention (2013) *Antibiotic resistance threats in the United States, 2013*, Atlanta: Centers for Disease Control and Prevention. Available at *http://www.cdc.gov/drugresistance/threat-report-2013/index.html*

Scientific method

Social scientific analysis of scientific method has a long history and a wide range of positions and perspectives. These three books cover the main modes of analysis. Chalmers considers the philosophy of science as applied to method, Fuller focuses on institutional and societal understandings of method and Harré looks at the role of experiments in history.

Chalmers, A.F. (2013) *What is this thing called science?*, Buckingham: Open University Press.

Fuller, S. (1997) *Science*, Buckingham: Open University Press.

Harré, R. (1981) *Great scientific experiments: twenty experiments that changed our view of the world*, Oxford: Oxford University Press.

Science and technology studies (STS)

Sismondo's book is a good overview of STS, and there is a comprehensive discussion of the emergence of laboratory studies in chapter 9:

Sismondo, S. (2009) *An introduction to science and technology studies*, Oxford: Blackwell.

The pioneer and originator of what has come to be called STS is Bruno Latour. This influential book presents his perspective on how to look at science and technology, as things that unfold through action:

Latour, B. (1987) *Science in action*, Cambridge, MA: Harvard University Press.

3

Scientific Knowledge

Cognition is the most socially-conditioned activity of man, and knowledge is the paramount social creation.
Ludwik Fleck, *Genesis and Development of a Scientific Fact* ([1935] 1979)

The previous chapter looked at the production of formal scientific knowledge in scientific workplaces. Experiments in laboratories are a clear example of the connection between action and formal knowledge: the actions of experimenters in laboratories result in the production of formal knowledge which appears as scientific papers in journals. Although experiments are not the only method by which formal scientific knowledge is generated, they are the underpinning of formal scientific method and are the core of much formal scientific endeavour. Formal scientific knowledge requires translation between the laboratory and the scientific paper, and subsequently a much greater degree of translation is required as formal scientific knowledge is disseminated in wider society, for example to social scientists or journalists. Moving away from experiments – that is, shifting the knowledge gained in a laboratory into other places, even just to other formal scientific locations such as specialist journals – means that there is a change in form and meaning. Scientific knowledge will thus take a wide range of different forms, will have a range of content, and will use a variety of different linguistic and discursive constructions.

All philosophical and social science accounts of scientific practice and scientific knowledge serve to create a picture of science in contemporary society: they represent science and scientific knowledge to us. Most philosophy and sociology of science considers scientific knowledge to be an object that is fairly standard with characteristics that are shared between all the different forms that it takes. In addition, many of the accounts of epistemology of science see scientific knowledge as being separate from, and often superior to, other forms of knowledge. This chapter will examine the relationship between the practical activities of scientists in laboratories and their production of scientific knowledge, the formal scientific knowledge of journals and the different ways that philosophers and sociologists of science make sense of this knowledge.

A major concern in this chapter is to challenge the standard account of science which sees scientific knowledge as better, bounded and clearly identifiable. The standard account is deeply embedded in our society, promoting scientific knowledge and method as having a status higher than other forms of knowledge and methods of inquiry. The standard account is also deeply embedded in the institutions of science itself, and the dominance of science in education results in a proliferation of this account. It is clearly visible too in many cultural representations of science.

Understanding how scientific knowledge achieves the status that it has in society requires a critical approach to such accounts. Where the first picture of science we examined was science taking place in the laboratory with its outcome of formal scientific knowledge, the second will be this: the picture of science created through formal appraisal of scientific knowledge. This way of considering scientific knowledge takes a variety of forms, but in general would be described as epistemology of science – the examination of the basis of scientific knowledge claims.

We have seen in the previous chapter a location and a set of processes that lead to the production of scientific knowledge. We can see a direct relationship between the results created in experiments and the narratives that emerge in scientific papers. We can also identify a range of social factors implicated in how experiments take place (e.g. team work, group discussion) and we can identify some external structural factors (e.g. funding regimes and the peer-review process) that affect what is done in the laboratory and thus what appears in publications. How can we start to make sense of these different elements such that we can understand why scientific knowledge emerging from the laboratory has the structure and character that it does? Two main approaches – sociology of science and philosophy of science – are available to us. However, we should also note that the discourses of science are constrained by tradition, a topic that is addressed by the history of science and which will be further investigated in chapter 4. In the following sections a brief overview of key themes in the sociology and philosophy of scientific knowledge will be given.

We will firstly examine the main trends in epistemology of science, providing an overview of the terrain. Then we will look at different sociological accounts of what is happening in the production of scientific knowledge. The aim here is to provide some background to answering the question 'how do philosophers and sociologists of science make sense of scientific knowledge today?' A secondary aim is to show that philosophy and sociology of science, even contemporary post-structural approaches, often serve to construct a picture of science, scientific knowledge and scientists that sees each in essentialist terms.

The standard account

What we will call the standard account is widely held, particularly amongst working scientists, and is a major component of the public image of science in contemporary society. It can be called the standard account of formal scientific knowledge when applied to single instances of the emergence of scientific knowledge (e.g. in an experiment), and the standard account of formal science when applied to the analysis of all formal science activities (e.g. when offering a description of the work of a scientific community). At this point we have to note that for the most part STS has departed from this account, seeing it as wholly inadequate to describe what is actually happening in the production of scientific knowledge. Yet it remains the case that a huge number of practitioners of scientific method still hold to this account. The standard account of formal science describes a project for science (the discovery of facts about the natural world), and prescribes a method – what we can call formal scientific method – by which this should be carried out. Formal scientific method relies, in experimental sciences (physics, chemistry, biochemistry, etc.), on the construction of experiments where hypotheses derived from theories are tested in conditions that allow for the strict control of all variables. In discovering sciences (e.g. botany, geology, astronomy), the equivalent of experiments can be seen in the construction of strict observational regimes where observational data is collected in precise ways, measurements are made in as accurate a manner as possible, and the theories generated by such a process are based on the collection of as many facts as possible.

The standard account proceeds from this premise:

Science is a form of knowledge that produces facts and fact-like statements.

This core tenet conflates science and knowledge, seeing them as indivisible. This means that the standard account is *not* defining science by its methods, locations, practitioners or shared value- or belief-systems. It defines science purely in terms of its goals and its outcomes. By this account, science is the knowledge that emerges from scientific activity, or even from non-scientific activity, if the knowledge conforms to stipulation of facts and fact-like statements. The definition is not about process, location or any other factor that may contain an allegedly 'subjective' element; just what the statements themselves look like. Finally, the word 'produces' contains some necessary ambiguity. The standard account of science assumes that the natural world is a passive, external object that objective science can confront in such a way that the underlying rules and laws of its operation can be discovered. Yet the standard account is forced to be vague about *how* these laws and rules emerge from formal scientific endeavour, as they are a product of human activity, not an artefact of nature (see box 3.1) that has been hidden for us to uncover.

Box 3.1 What is nature?

Central to the project of formal science is the understanding of 'nature' – hence the term 'natural sciences'. This may seem obvious to us, but we need to be wary of simply taking our current, dominant understanding of what science does and assuming it is an axiom. If we take a step back and consider what it is that we actually mean by 'nature' then some problems present themselves.

Firstly, 'nature' – the word as deployed in our language inside our form of life – takes its meaning from its use. Think about the differences between how we understand and deploy phrases such as 'human nature', 'nature versus nurture', 'natural history', 'the science of nature', 'natural law', 'natural yogurt'. The vocabulary we use that deploys 'nature' is often exoteric in its form, in contrast to the esoteric vocabularies of formal science where 'nature' is used quite differently (say in addressing the 'nature' of a cell, molecule or galaxy). This is a manifestation of the 'exoteric paradox', where the more a scientific community tries to make itself understood in exoteric terms the further it removes itself from its esoteric vocabulary and the more it gets involved in trying to resolve ambiguities in these big exoteric expressions (Baldamus [1979] 2010b).

Secondly, we need to think about how nature is possible; if we see it as simply the thing that is 'out there', beyond the self, then we need to ask and answer the tricky question of how it is that some things are defined as 'natural' and others not.

For Kant, who was a practising formal scientist and whose discoveries in astronomy are still central to our understanding of the formation of stars and planets and of the multi-galaxy universe (Kant 1969), nature was nothing but the *representation* of nature, and the unity of nature was provided by the conscious processes and perceptions of observers. Subjects synthesize observed elements into nature. Sociologist Georg Simmel (1858–1918) used this Kantian point of departure to offer a contrast between society and nature. Where nature is made possible by being actively synthesized by observers, society pre-exists the observer and is already synthesized into a whole (Simmel 1963).

This perspective is echoed in the relativist science studies of the 1970s onwards which considered nature simply did not exist, but was rather called into being by scientists acting, talking, interacting with non-human objects and writing together (Collins 1981).

More recently Bruno Latour's actor-network theory (discussed later in this chapter), departing from the Kantian perspective, proposes that we should wholly dispense with the concept of 'nature' – 'conceived as the gathering of all non-social matters of fact' – as an unhelpful idea. '"Society" and "Nature" do not describe domains of reality, but are two *collectors* that were invented together, largely for polemical reasons, in the 17th century' (Latour 2005: 109–10).

The core tenet in one way does describe what scientists in laboratories are doing – the end result of their endeavours **is** the production of facts and fact-like statements (although there can be much disagreement about what a 'fact' actually is). But the production of facts and fact-like statements does not necessarily distinguish science from non-science. By this criterion many things we do in everyday life would count as science. As I look at my desk I can see my laptop (a fact) and I can say 'the laptop is in front of me', which is a fact-like statement. But this isn't science (although it could be construed that I am deploying a form of rational explanation akin to scientific rationality). To clarify the distinction between science and non-science, the standard account proceeds to qualify and extend its description of science. The key elements of this extended description are as follows:

- Science is not metaphysics, where metaphysics is a range of sweeping generalizations. Science is a series of factual propositions. These factual propositions are connected to each other by a common subject matter (the natural world) and a project (the extension and completion of knowledge of the natural world).
- Science connects factual propositions through the use of theories. Scientific theories describe what is known about the world and extends science to make predictions about what is not known about the world.
- Science is empirical. It is based upon experience, i.e. actually perceiving things rather than just creating theories about them.
- Scientific knowledge is applied rationality: it is produced using concrete and rigorous methods. Scientific knowledge relies on the use of scientific rationality.

- Scientific knowledge is a direct refutation of religious experience. Religion relies upon faith, science upon facts and rationality.
- Scientific knowledge is based upon objectivity and seeks to remove subjectivity from analysis of the world.
- Scientific knowledge has definite outcomes; it makes science 'work'. It transforms our lives through producing technological breakthroughs, cures for illnesses, new ways of understanding our environment, etc.
- Scientific knowledge is cumulative and progressive, through theories coming together and supporting each other. We know more about the natural world than we did in the past and we build our knowledge on foundations that have been laid down in the past.

These elements of the standard account of science emerged in the centuries following Newton's unification of the laws of physics. There is an orthodoxy surrounding these principles. This strong version of science is exemplified and expressed most clearly by adherents of the Vienna Circle, the influential philosophy group that began meeting in the 1920s and included Rudolf Carnap, Viktor Kraft and Moritz Schlick. Their logical positivism relied on an understanding of science that considered scientific knowledge (and mathematics) to be better, more robust and more valid than other forms of knowledge, largely because it constructs theories that describe the world. The Vienna Circle's ideas have a powerful force even today, and many scientists when asked to describe their understanding of science will offer explanations that partially concur with the Circle's. Its logical positivists thought that theories attain their scientific status through *induction* (the drawing of inferences from a number of particular cases) and *verification*. Verification argues that a theory is proposed and that it makes predictions that can be tested through observation. Scientists will sceptically adopt a theory, and will then test it by making many observations. As the observations that confirm the theory accumulate, it achieves a scientific status. Therefore a proposition is scientific to the extent that it can be verified by observations, and a scientific theory becomes 'true' when a sufficient number of observations or experiments have been accumulated.

However, an associate of the Vienna Circle, but a dissenting voice, Karl Popper (1902–94), is also influential in the contemporary construction of the standard account. Popper and Carnap appear to be quite opposed about some aspects of formal scientific work, but, as Ian Hacking points out, '[t]hey disagreed about much, but only because they agreed on basics' (Hacking 1983: 3). Popper's theory of the logic of scientific theories both refutes and extends the ideas of the Vienna Circle. It refutes them in that their theory of verification is disproved by Popper, but extends them in that Popper's defence of science strengthened the logical positivists' arguments that scientific knowledge is superior to non-scientific knowledge. Both Popper and the Vienna Circle agreed that understanding the role of theory is central to this endeavour. 'The empirical sciences are systems of theories. The logic of scientific knowledge can therefore be described as a theory of theories' (Popper 2002: 37).

Popper's understanding of the formation of scientific knowledge rests upon an understanding of theory that was totally new at the time of his writing (*The logic of scientific discovery* was first published in 1935). Popper's refutation of the theory of verification was quite devastating. His theory of falsification (as opposed to verification) suggests a quite different process of theory construction for science. Theories arise in response to pressing problems – they emerge as attempts to solve problems in already existing theory. The new theory will be tested in the same way that the logical positivists suggest, that is, through naming observations and carrying out experiments. But according to Popper, scientists aren't carrying out observations and experiments to *verify* the theory; rather they are doing this in an attempt to *falsify* the theory. No scientific theory can ever be proved true, but all scientific theories can be proved false; just because scientists have collected many observations that confirm the theory doesn't mean the theory is correct, just that evidence to the contrary has not yet been found.

Popper's account became the dominant story of how formal science was taking place. It describes the actions of scientists in laboratories, seeing them as engaged in efforts to falsify existing theory as a way of progressing scientific knowledge: theories that could not be falsified became 'truthful', although according to Popper's account, scientists would retain a degree of scepticism about such theories. Similarly, Popper's account describes how the project of science proceeds, through collecting and accumulating different 'truthful' theories that become increasingly reliable. Finally, this account describes why scientific knowledge is 'better' than other sorts of knowledge: because it emerges from theories that are falsifiable. Popper's theory is specifically aimed at validating scientific knowledge, but it is also a clear critique of non-scientific knowledge and theories. He famously put it to work against Marxism (Popper 1945) to refute the latter's claims to being a 'science of society'.

This version of scientific knowledge is an insider's account of science. Writing in the 1930s against a political background of competing and murderous ideologies, Popper offered an account of scientific knowledge as a better way of making sense of the world than ideology. His theory proposed that humans could find rational and objective explanations for the world that would not be subject to ideological influence and could be a force for general good. Popper and the Vienna

Circle equated science with applied rationality, and science and rationality became, for a time at least, inextricably linked.

Scientific realism

Philosophy of science for a long time offered two separate positions with which to make sense of the objects that appear in scientific experiments: realism and anti-realism. In recent years, a third position of social constructionism has grown in prominence. We'll examine each, but should note at the outset that the standard account relies on an ontological and epistemological perspective which makes precise and specific claims about the world and what we can know about it. Realism is 'the default, commonsense position' (Bortolotti 2008: 96) which assumes that there are objects in the world independent of our minds, and which our perceptual experience can access. Ian Hacking notes that some philosophers of science have found problems with using the term 'realism', so he uses another name: inherent-structurism. 'I suppose that most scientists believe that the world comes with an inherent structure, which it is their task to discover' (Hacking 1999: 84). Paul Feyerabend agrees with this analysis, although he extends his assessment, considering scientists to have adopted an 'ideology' of materialism that they use to describe a 'depressing reality' of a world without purpose (Feyerabend 2011: 35).

These ontological positions address the problem of whether or not the objects that are described by scientific theories and investigated by scientific experiments really exist. This is not as strange a proposition as may be first thought. Some of the objects that scientists are concerned with, particularly in the discovering sciences, are obviously 'there' and tangible to the senses: planets, glaciers, human beings. But many objects, particularly in the experimental sciences, are not perceptible to the naked human senses (for example atoms, molecules, cosmic rays) and may only be perceptible through inferential (i.e. indirect) procedures. Given that the standard account of science places huge store on an empirical method, i.e. a method based upon experience and perception, this can cause some problems. If the experimenter cannot actually see the phenomena in question, are those things really there as described by the theory, or is it just a coincidence that the theory is predicting results that are being achieved? This is the core of the disagreement between scientific realists and anti-realists:

> Scientific realism is the position that scientific theory construction aims to give us a literally true story of what the world is like, and that acceptance of a scientific theory involves the belief that it is true. Accordingly,

anti-realism is a position according to which the aim of science can be well served without giving such a literally true story, and acceptance of a theory may properly involve something less (or other) than belief that it is true. (van Frassen 1980: 9)

Scientific realism works like this:

(1) We have theories that are correct, theories that we have worked out from extensive experiments and from our current paradigm.
(2) Because the theory is correct, the objects that the theory names and manipulates are real. They actually exist. For example, the theory of vulcanology identifies volcanoes as actually existing, and the theory of electricity names electrons as actually existing.
(3) Until our theory is proved wrong we can assume that the objects are actually there and treat them as being real rather than as being thought constructs, concepts or possibilities.
(4) A major goal of science has always been the construction of correct theories – therefore we do not need to change what we do.

> Scientific realism says that entities, states and processes described by correct theories really do exist. Protons, photons, fields of force, and black holes are as real as toe-nails, turbines, eddies in a stream and volcanoes. The weak interactions of those particles are as real as falling in love. Theories about the structure of molecules that carry genetic codes are either true or false, and a genuinely correct theory would be a true one. (Hacking 1983: 21)

Anti-realism says the opposite to realism. There are no such things as electrons. There is, of course, electricity, but our theories are essentially fictions that we use to help us predict and produce events that interest us. Electrons are fictions by this account. There is no necessary reason to believe that just because there is electricity and we can produce an account that adequately explains electricity, through making up a fictional particle called the electron, we have somehow produced a correct theory. In Hacking's terms; turbines, yes, electrons, no (ibid.). Just because we can make models of the world does not mean that that is actually how the world really is.

Anti-realism is a more difficult doctrine to assume. Our dominant culture 'believes' in realism – the common-sense default: the representations of science that surround us reconfirm this realism. Pictures of DNA molecules that, for example, illustrate the achievement of the Human Genome Project look as if they are concrete objects that are 'really there'. Anti-realists would not try to deny that there are inheritance factors that work in passing on genetic information, but they would say that there may not actually be specific molecules of DNA that really look like the ones in the newspaper pictures. For anti-realists the theory of DNA is a model that helps us to organize ideas and data: it is not a literal picture of how things actually are (Hacking 1983: 22). Anti-realists do not dispute the results of experiments, or even the status of science, but instead dispute the form that theories actually have.

Van Frassen extends this position, and suggests that anti-realists can operationalize their experimental work as constructive empiricists: 'Science aims to give us theories that are empirically adequate; and acceptance of a theory involves a belief only that it is empirically adequate' (van Frassen 1980: 12). From this anti-realist position all these constructive empirical accounts must mesh together such that all the phenomena they describe fit into the same model. For van Frassen the achievement of such a model is described by the current situation of scientists being committed to their theories.

Problems with realism

Scientific realism is strongly adhered to by many scientists and philosophers of science; it seems to make sense in their day-to-day activities. But there is a problem at the core of realism. Simply stated, there is no necessary step from accepting that entities exist to believing that they exist. In addition, if we look at the history of science and assume, plausibly, that scientists believed that their theories and the entities that their theories described were correct, how do we justify a realist perspective when we know that these theories have been proved to be false and the entities they described did not exist? A good example is the late nineteenth-century theory of the aether. Newtonian science unified the laws of physics and posited a series of principles by which light and energy were propagated. Central to this was the idea that electromagnetic forces such as light are propagated through some kind of medium, in the same way that the energy of waves is propagated through water. Physicists proposed that the universe was filled with a substance that acted as this transmission medium – the aether – and that stars and planets were surrounded by it. In the late nineteenth century a famous experiment – the Michelson-Morley experiment – found no evidence for the existence of the aether. This plunged physics into a crisis situation: a contradiction between theories and the results of experimental work. This crisis was resolved by the emergence of the new Einsteinian paradigm that dispensed with the idea of the aether altogether and saw space as being a vacuum through which light was propagated.

The problem here is referred to as 'pessimistic meta-induction', made famous by philosopher Hilary Putnam (1977). If theories are considered to be true now, and previous theories were considered to be true in the past but subsequently proved false, how can we justify our belief that current theories are true? The idea that just because a theory is empirically successful it must be true is hard to sustain, given that the history of science is full of examples where theories are subsequently proved to be incorrect. 'On the basis of evidence from history of science, we might come to believe that the theories we currently accept will be regarded as false in the future and replaced by other theories, in spite of the fact that they are empirically successful right now' (Bortolotti 2008: 99).

Can anti-realism address this crisis? Anti-realists assuming a constructive empiricist perspective see scientific theories as just ordering tools that allow scientists to formulate clear questions about the natural world. Theories from this perspective are 'empirically adequate' but not necessarily true (Couvalis 1997: 174). Choosing between theories is a pragmatic task: which theory fits our observations of the world best. But this still leaves us with a major problem: the ontological status of the natural world is not considered by the anti-realist/constructive empiricist position. Rather, this analysis is at the level of the practices of the scientist and the conventions of the scientific community. Ultimately anti-realism cannot come to a decision on whether the objects that are described by theory really exist or not, and this is an uncomfortable position to hold. Why? Because some objects that scientific theories describe clearly do exist (volcanoes) whilst others may or may not (dark matter): science is a universal endeavour, according to the standard account, and an ontological perspective that applies to only some parts of the natural world is unacceptable.

However, there are problems with both of these positions when we consider wider, social issues. Whilst one critique of the anti-realist position is that it is in part a social and psychological theory (it makes assumptions about what the scientist is thinking), both of the theories assume that language is a fixed and determinate system which links words to objects in the world. This is certainly at the heart of Putnam's revised version of realism (internal realism) (Putnam 1977: 488–9). But words are simply not like that and although we can see scientific discourse as an attempt to get words to have more precise and fixed meanings, there is always an element of contingency in the use of words, even inside the laboratory or the formal science paper. Neither realism nor anti-realism can escape from a form of determinate language that is not congruent with how language works in the social world: it assumes that the referent 'electron' for both realists and anti-realists has a fixed meaning that is held by all members of a scientific community and which is immutable inside a theory. A social understanding of language and meaning such as Wittgenstein's tells us that this is wrong. A social analysis of laboratory spaces and the discussions held by scientists further confirms this, as Fleck noted a considerable time ago:

Thoughts pass from one individual to another, each time a little transformed, for each individual can attach to them somewhat different
associations. Strictly speaking, the receiver never understands the
thought exactly in the way that the transmitter intended it to be understood. After a series of such encounters, practically nothing is left of the
original content. Whose thought is it that continues to circulate? It is one
that obviously belongs not to any single individual but to the collective.
(Fleck [1935] 1979: 42)

In contrast, some philosophers and sociologists of science have
attempted to come to terms with the contingency of language and the
vagaries of realism. The constellation of theories that come together
under the general heading of 'social constructionism' all have a common
root in an abiding concern with language and, often, a Wittgensteinian
influence. To investigate social constructionism we need to look in a bit
more detail at Wittgenstein's work.

Wittgenstein, language and science

The later philosophy of Ludwig Wittgenstein fundamentally altered
the philosophical analysis of language, and the impact of his work can
be seen throughout contemporary thought, particularly those forms
that describe themselves as being 'anti-foundational', 'post-structural'
and/or postmodern. Wittgenstein's work is written in a deceptively
simple style, which makes it easy to read and understand, but hard
to explain or replicate. Wittgenstein did not explicitly discuss natural
science often, so much of what is presented in what follows extrapolates from his areas of study.

Wittgenstein's philosophy shows that there are no absolute fixed
meanings attached to words. He provides us with a dictum – 'the
meaning of a word is its use in the language' – to remind us of this
(Wittgenstein [1953] 1958: §43). Further, Wittgenstein says that understanding a language means understanding a form of life (ibid.: §19).
Thirdly, he notes that there can be no such thing as a 'private language' –
all language is shared and relies on this shared aspect for meaning to
be intelligible (ibid.: §243). These basic principles of Wittgensteinian
philosophy have major implications for our understanding of science
and scientific knowledge.

To start with, the meanings of words are contingent upon their use
in language. This means that the meanings of words shift with time
and context. Regardless of the fixity of the external world of objects,
Wittgenstein is saying that our language will never have fixed points,
is never nailed down to the objects it is describing. This creates problems for formal science. Imagine experimenters explaining their work
to each other. They appear to be using a number of highly specialized

terms to describe things in front of them. Yet these terms rely upon our everyday language for their meaning. Whilst it is possible to see that much of science is concerned with defining objects in tight and rigorous ways, this can never be achieved fully. There will always be some 'looseness' around the meanings that are attached to words, even to highly specialized words such as 'polymerase chain reaction'. This can be seen quite clearly if we think about how words, even 'tightly defined' scientific terms, change their meaning through time.

If the meanings of words are contingent upon their use, we will have great problems in offering descriptions of reality if we try to adhere to the 'truth' and 'fact' orientation of the project of science. As the meaning of words is internal to the language-game being enacted at any one time, reference to reality or the external world has to gain its meaning from a specific context in a particular human practice (Trigg 1993: 29), not a universal and immutable context and set of practices. Everything thus becomes dependent upon our frame of reference: we can have no external 'god's-eye' view because if we were to do so we would need a new, separate and 'private' language.

We cannot do this because our reality is circumscribed by what we can describe using our current language. Constructing new linguistic expressions as an attempt to 'firm things up' cannot work because those linguistic expressions also rely upon our current language for their meaning. The limits of our world are the limits of the language that we use: changing the language simply changes the limits, not the location from which we are observing the world or describing it.

In terms of the language that science uses, it cannot have its own special and separate language; it will always rely upon everyday language to construct meaning. Scientific language may look as though it is 'different' from everyday language, but fundamentally it isn't (Erickson 2004).

Implications for looking at science

This is quite clear when we look at 'translations' taking place between different parts of science as a whole. Inside the lab we can see a group of laboratory workers together talking about what they are doing and seeing. They describe the world in front of them in mundane and everyday terms for the most part, occasionally moving to specialist descriptions to make sure co-workers are identifying the same objects that they are. However, the formal scientific paper that is written from these activities uses a discourse that is much more rigorous – effectively a translation of the work of experimenters into the formal scientific discourse of the paper. Subsequently we can see further translations from the formal scientific discourse of a paper into, say, a newspaper or media account where specialist terms are abandoned in favour of 'easier' ones. All three forms of discourse are intelligible to observers outside of the thought community that did the experimental work, and all three discourses are describing the same objects. Different discourses apply in different thought communities, and these discourses will reflect different thought styles and meanings attached to words. Formal scientific discourse may be a 'dialect' in the same way that youth argot may be a dialect, but it is not and cannot be a separate language.

The basic principles of science cannot be given rational justification by Wittgenstein. Ultimately everything comes down to human practices and the construction of systems of belief inside thought communities. There is no correspondence theory of truth – this theory, this fact. This is not to say that Wittgenstein is anti-science. However, rationality cannot, in this perspective, have anything transcendentally special about it. It is just another belief and practice like any other, albeit one that has a lot of power attached to it and one that does produce real results in terms of organizing our practical investigations of the world. Rationality is a language construct like any other concept.

From this perspective science is no longer the privileged location of the truth, just another system of signification. Indeed, we can use Wittgenstein's theory to begin to treat science as being similar to a belief-system. Science generates its own rule-governed activities, and its own meanings in its language-games. We may want to say that the language-games of science are more tightly ordered than the narratives that we use in everyday life. But science still relies upon everyday language and narratives to provide meaning and make sense. It is, at the very least, inextricably linked to society.

Challenges to the standard account

For a long time sociology of science was not concerned with what took place inside laboratories in terms of how experiments were carried out. Practical activities and the relationship between human actors and non-human objects were not a concern for sociology of science until the late 1970s. Prior to this point, sociology of science – which has a long tradition of inquiry as a sub-discipline of sociology – had been concerned largely with the institutional structure of science and the scientific community, and with issues of status ascription inside the scientific community. Classic studies in sociology of science, the most influential of which was Robert Merton's delineation of the institutional imperatives of science (see box 5.1 on p. 142), focused on how science as a whole comes together as a set of institutions, why it is superior to non-science, and how science is integrated into the wider social structure. Similarly, Hagstrom's analysis of the scientific community looked at the scientific community as a whole, how it circulated ideas internally and disseminated ideas externally (Hagstrom 1965). Merton's work on the scientific community will be discussed more fully in chapter 5, but at this point it is worth noting that such studies of scientific communities were common in the 1960s and 1970s, and all took a similar approach, i.e. they did not question the form, content or status of scientific knowledge (a topic considered to be the preserve of philosophers of science) but focused on institutional aspects of the human organization of scientific activity.

Even Max Weber's critical gaze ignored the status of scientific knowledge, although his essay 'Science as a Vocation' was a highly influential and instructive text (Weber [1918] 1989; Erickson 2002), not least for Merton. The themes of Weber's essay are still of central importance to sociology of science today. Weber does three things in his essay. Firstly, he describes the external conditions of science as a vocation, primarily concentrating on the institutional organization of science in contemporary Germany and offering some comparisons to the situation in North America. Weber focuses on the rationalization and institutionalization of science, justifying this in a somewhat self-deprecating way as the 'pedantic custom' of political economists such as himself. He also describes the entry route and career path of recruits to the world of academic science, and notes the proletarianization of the intellectual through being transformed into a 'specialist' (Turner 1992: 100). Secondly, he goes on to discuss the inward calling for science that scientists possess, noting that this is what his audience really want him to speak about. Here his focus is on the motivation towards a scientific career that scientists share. Finally, he offers a discussion of what science actually is, and what role it fulfils in society (Erickson 2002: 30).

Standard textbooks such as Barber and Hirsch's *The Sociology of Science* (1962) simply did not question what was happening inside laboratories other than pointing out that experimental method had to conform to the norms of the scientific community, and that deviation from such norms would dramatically reduce the status and quality of scientific knowledge, turning it into 'non-scientific' knowledge, a pretty basic calculus. All of these approaches reinforced an essentialist conception of science, scientific knowledge and scientists. Such sociology of science sees science as being an institution that is in some way separate from, although connected to, the rest of society. It sees scientific knowledge as having an intrinsic superior quality to other forms of knowledge due to its method of production (the scientific method) and its embodiment of rationality and truth. It also sees scientists as being separate and different from other members of society, at least whilst they are carrying out their experiments in the laboratory. Unlike non-scientists, who are using rough-and-ready forms of practical reasoning to make sense of their world, to produce knowledge and to complete tasks, scientists, from this perspective, are producing knowledge by using a form of purer rationality that allows their results to have a closer connection to truth.

It is interesting that this kind of sociology of science is a reflection of the general societal view. In Fleck's terms we can see the importation of exoteric ideas into the esoteric thought community of the sociology of science. This thought community clearly modelled itself on the thought style and norms of the esoteric formal science communities it was considering. We can see a quite definite starting point to the changes in sociology of science towards a much more critical approach to science in the late 1960s. The key text that marks this shift is Hilary Rose and Steven Rose's *Science and Society* (1969), a trenchant examination of British science policy in the post-war years. Rose and Rose found that far from being a neutral and separate institution science was the product of certain ideologies, philosophies and economic and political structures, and was largely closed to public scrutiny, none of which was a good thing. Rose and Rose's book was published at a time of great social and political upheaval in Western capitalist societies, events that were later dubbed the 'world revolution' by sociologist Immanuel Wallerstein, whose work we will discuss in chapter 4.

In the late 1970s a new trend in sociology of science emerged, a project that is widely referred to as the sociology of scientific knowledge (SSK), which employs a form of social constructionism as its main mode of theorizing the social world. This project has a history that can be traced back to the roots of sociology and sociological analysis, but the term 'social construction' was imported directly from Peter Berger and Thomas Luckmann's *The Social Construction of Reality* (1967), an influential phenomenological approach to making sense of the social world. SSK's first significant manifestation was in the form of the 'strong programme', an attempt to understand the content of scientific knowledge in sociological terms. The strong programme originated in the sociology of knowledge produced by Barry Barnes and David Bloor, both of whom were working at Edinburgh University (Barnes and Bloor 1982). The strong programme also appeared in the work of Harry Collins in the late 1970s, and is associated with him and with other social investigators of science who were interested in SSK and, more recently, the general field called STS (both of which are discussed in more detail in what follows). The relativism of the strong programme has been influential on much of science studies since the 1980s. SSK also relied upon earlier work that prepared the ground specifically in the area of science, particularly Thomas Kuhn's history of science (especially Kuhn [1962] 1970), and later Paul Feyerabend's 'anarchist' analysis of scientific method (Feyerabend 1978a, 1978b).

Between them, these relativist approaches, in sociology, history of science and philosophy, challenged the standard account of scientific method, scientific knowledge and science as a project. Significantly, they gained prominence at roughly the same time – the late 1960s and 1970s – and began to converge around a set of ideas that described how scientists were involved in constructing, not discovering or finding, scientific knowledge. The relativist challenge to the standard account of science made its mark not by challenging head-on the institutional aspects of science and the scientific community discussed by sociologists of science such as Hagstrom, Barber and Merton, but rather by looking directly into the places where knowledge was produced and challenging the alleged objectivity and formality of scientific practice and experimental method. Overall a convergence in sociology and philosophy of science takes place with both disciplines focusing on the status and nature of scientific knowledge, and on how scientific knowledge is validated. We have already seen how this was connected to the focus on language that Wittgenstein's philosophy highlighted, and the social constructionist movement that emerged in the late 1970s owed much to this, and to the other direct sociological descendants of Wittgenstein, the ethnomethodologists. The starting point for the emergence and continued hegemony of social constructionism, and its current main theoretical form in actor-network theory, was the SSK movement of the 1970s. We will first consider social constructionism and then look at how Ludwik Fleck's 1930s work presaged and predicted much of this.

The emergence of social constructionism and actor-network theory

The emergence of SSK in the 1970s and 1980s started from some fairly basic, but radical, premises: perhaps science is just a practical activity – a form of social interaction and work that happens to have a rather special product (facts/truth). From this perspective we could treat scientists in much the same way as any other social group, and we can look at how such a social group is building social order and meaning around itself. But this is, potentially, a threatening route for the project of science: from here it was only a small step to realizing that as well as creating order in terms of social interaction, etc., scientists were also constructing the 'truth' and their 'facts'. These ideas are at the core of the social constructionist approach to science.

Sociology of science, in the years following Kuhn's publication of *Structure of Scientific Revolutions*, gradually moved away from a concern with the institutions that scientists inhabit (e.g. scientific communities) to being concerned with the activities and interpretations of scientists themselves. This meant a sharp change of methods. Where sociology of science had been able to 'look in' from the outside to make sense of scientific knowledge and its relation to institutions, this new approach required sociologists to observe the actual day-to-day processes of doing science to begin to construct accounts of what was happening inside science.

SSK allied itself to the strong programme. Collins's 'Empirical Programme of Relativism' (Collins 1981) provided a formulation of how social constructionist studies should be taken forward. Observation should be made of how it is that scientists themselves offer different interpretations of reality, or nature: 'the truth or falsity of scientific findings is rendered as an achievement of scientists rather than of Nature' (Pinch 1986: 20).

Trevor Pinch, a close collaborator of Collins for many years, offers a clear definition of what the strong programme in science studies entails.

> Let me first state the most important thesis which guides this work and other 'social constructivist' studies:
>   In providing an explanation of the development of scientific knowledge, the sociologist should attempt to explain adherence to all beliefs about the natural world, whether perceived to be true or false, in a similar way.
>   This thesis, which is sometimes referred to as the principle of symmetry or equivalence within the so-called 'Strong Programme' in the sociology of knowledge, simply sets out the widest possible terrain for sociological explanation. (Pinch 1986: 3)

There are major implications of the strong programme. By adopting this frame of reference investigators must treat all true and false beliefs in exactly the same way and use similar modes of explanation for them. This results in a situation where, inevitably, a number of different versions of knowledge will emerge, all of which will have validity:

> If true and false beliefs are to be treated in the same way, then we are not merely talking about distortions of the picture of knowledge which is painted. What is being claimed is that *many pictures* can be painted, and furthermore, that the sociologist cannot say that any picture is a better representation of Nature than any other. Scientists can socially construct many different versions of the natural world. (ibid.: 8)

Science, even the formal science of the laboratory, becomes, from this perspective, relative, with meaning contingent upon how interactants in a laboratory are collectively 'making up' the reality that surrounds them. Scientific knowledge is relative. The strong programme assumes that truth and falsity are constructions with no necessary connection to 'reality'.

> The importance of the philosophical arguments about relativism in the 1970s was, in retrospect, not that they showed that relativism was true but that it was tenable and therefore could be used as a methodology for the study of science. (Collins and Yearley 1992: 303)

The first major account that did this was Latour and Woolgar's *Laboratory Life* (Latour and Woolgar 1979), a meticulous ethnographic analysis of work, interactions, actions and meaning constructions in one biochemistry laboratory. The relativist approach in the sociology of science suffered from a major problem, namely that showing relativism is difficult and open to interpretation. Latour and Woolgar examined in detail what exactly was taking place *inside* a biochemistry laboratory and provided empirical data that showed relativism at work in a setting where it was not expected to be found. Their work acted as a starting point for a great many other social constructionist laboratory-based studies.

Latour and Woolgar observed daily life in a high-profile and prestigious biochemistry laboratory (the experiments they observed were subsequently part of a Nobel Prize win) from an anthropological and highly sceptical perspective. They took nothing at face value, choosing, for example, to see the laboratory not as a workplace oriented towards the search for truth, but as a 'factory' that is designed to generate paper products (articles in scientific journals) from inputs of animals, chemicals and energy (Latour and Woolgar 1979: 46). They called into question the status of the humans and non-humans inside the laboratory, suggesting that non-humans could have a meaningful part to play in the generation of scientific knowledge. But most crucially of all they did not take at face value the core beliefs of the scientists, choosing to redescribe these in terms that implied a very different version of reality.

Rather than seeing scientists uncovering truths in nature through experiment and analysis of results, Latour and Woolgar found scientists constructing facts and generating a discourse of truth, through their practical activity, their interactions and actions.

> Scientific activity is not 'about nature', it is a fierce fight to *construct* reality. The *laboratory* is the workplace, and the set of productive forces, which makes construction possible. (Latour and Woolgar 1979: 243)

Categories of 'objective' and 'subjective' take on quite different meanings here, and Latour and Woolgar imply that an 'objective' approach is simply not possible: this is a core tenet of the relativist approach. Latour and Woolgar's perspective effectively abolishes the concept of 'nature', the substrate that scientists are supposedly working upon and the 'object' that the project of formal science is seeking to understand perfectly, and replaces it with a discursive construction:

> If facts are constructed through operations designed to effect the dropping of modalities which qualify a given statement, and, more importantly, if reality is the consequence rather than the cause of this construction, this means that a scientist's activity is directed, not toward 'reality', but toward these operations on statements. (Latour and Woolgar 1979: 237)

Collins expresses this idea even more succinctly: '[T]he natural world has a small or nonexistent role in the construction of scientific knowledge' (Collins 1981: 3).

Latour and Woolgar's work can be seen as a strong challenge to the standard account of the production of scientific knowledge. But it can also be read as an attempt to oppose technological determinism and cultural imperialism. Social constructionists in this mould argue that all discourses should get equal weighting when social scientists encounter and analyse them, and that there is nothing essential in science that means it is better than any other form of discourse. The social constructionist line suggests that the things around us and the things we talk about are created by our speech, and thus we have power to change them.

This radical approach appears to present an entirely relativist version of science. It is clearly not a realist approach, nor is it antirealist. Indeed, Ian Hacking dubbed the new social constructionist approach as 'irrealist' to mark its departure from previous models (Hacking 1988). Studies in the strong programme tradition often explain extremely well how it is that a group of laboratory workers will manage to achieve consensus on the objects that have emerged from experiments or that are surrounding them. However, it is noteworthy that these studies of formal science largely confine themselves to looking at scientists' interpretations of science, or looking at direct relationships between scientists and non-scientists. Such social studies of science themselves are still involved in maintaining a separation between science and society, maintaining a boundary between where science 'really is' (in the laboratory or the head of a scientific expert) and where science isn't (everywhere else, but notably absent from the minds of lay people and popular culture). Although expressing a version of anti-essentialism with respect to concepts, social theories and methodologies, these social constructionist accounts often end up reproducing an essentialist account of science through a kind of mimesis. They are taking the language and discourse that is creating one phenomenon (scientific knowledge) and are translating it into the language of SSK to create another phenomenon (the social constructionist account of scientific knowledge). This may be a process of clarification, of explaining what is happening, but it isn't a process that brings out key features of *why* things are happening in the way that they do.

We can see from the preceding that challenges to the standard account have led to the emergence of a discourse of relativism. Science is decentred and the essentialism implied by the standard account is refuted. In general these discourses become a post-positivist critique of science.

However, post-positivism is a diverse discourse, and although social studies of science have tended to adopt some aspects of post-positivism, what we can see is a proliferation of different approaches to making sense of science and technology. The emergence of STS in recent years has been characterized by a range of attitudes, political and theoretical, and a range of sites for research. Simply looking inside laboratories and other sites of formal knowledge has ceased to be the main location for STS, and new imperatives, particularly concerning the relationship between human and non-human (in the form of technological objects) actors have come to the fore. This is especially clear in the dominant orientation within STS, actor-network theory, which emerged in the 1980s and 1990s.

Actor-network theory

Originating in the work of Bruno Latour (1987, 1996, 1999, 2005), Michel Callon (1986) and John Law (1986, 1991, 1994), actor-network theory (ANT) is now a heterogeneous set of approaches that coalesce around a post-positivist, post-structural perspective that takes seriously the relationships between humans and non-humans. ANT's roots in SSK and social constructionism are clear. ANT is also the site from which a sustained analysis of technoscience emerged, as it argues strongly that 'technology' and 'science' are inextricably linked and cannot be separated. ANT starts from the point of challenging the general social theoretical idea that human social relations 'were simply unmediated relationships between naked human beings, rather than being made possible and stable by artefacts and technologies' (MacKenzie 1998: 14).

In marked contrast to earlier sociology of science, ANT sought to collapse the divide between the production of science and its use in society. To do this ANT proposes seeing the parties involved in the construction of science as taking positions in a network that has specific outcomes. These nodes in a network are 'actors' (humans) and 'actants' (non-humans); all can have similar agency and all are involved in construction of each other and the network. The process of constructing these networks is what is of interest to ANT, and the network itself is responsible for constructing the boundaries and categories of the world – nature, the social, politics, science, technology. For example, a biochemist working on a specific biological molecule will enrol other biochemists, funding agencies, actants (PCR machines, spectrophotometers, bacteria) to construct a solid network that can produce theories, experimental results, technologies. Subsequent to carrying out experiments further actors and actants can be enrolled (or they and/or existing ones can be disenrolled) in the network – the media, for example, can be enrolled to disseminate the experimental findings or to display technological outcomes. ANT relies on the principle of symmetry – there is no purpose in trying to identify the 'truthfulness' of what emerges from the network. Instead, ANT encourages the researcher to investigate what actors are actually doing rather than examining the product of the endeavours of scientists. ANT is thus radically different from previous ways of 'doing' sociology of science or STS (and ANT is rapidly expanding into other areas of social inquiry). It allows users to see the social world, or parts of it, as an assemblage of networks, each of which is composed of translations between co-existing mediators (Latour 2005: 108). The result is a rejection of previous categories and understandings of 'the social' in favour of a rather different perspective: 'I can now state the aim of this sociology of associations more precisely: there is no society, no social realm, and no social ties, *but there exist translations between mediators that may generate traceable associations*' (Latour 2005: 108).

To achieve its aims, ANT proceeds through identifying case studies of interest, and then investigating the network of relations that emerge from given situations. Latour's examination of a failed transport initiative (project Aramis) in France from the 1960s to the 1990s identifies a complex of network relations, and relations between humans and non-humans (Latour 1996). He uses primary interview material coupled to original documents to chart the rise of project Aramis – noting who the key actors are, and letting them speak. For Latour, non-human objects can also tell a story:

> machines . . . are cultural objects worthy of . . . attention and respect. [Humanists will] find that if they add interpretation of machines to interpretations of texts, their culture will not fall to pieces; instead it will take on added density. I have sought to show technicians that they cannot even conceive of a technological object without taking into account the mass of human beings with all their passions and politics and pitiful calculations, and that by becoming good sociologists and good humanists they can become better engineers and better-informed decision makers. (Latour 1996: viii)

As the narrative progresses it becomes clear that the engineering aspects of the project are intertwined with the human interests of the human actors – and are sometimes displaced by these. Latour argues that meetings of various spokespersons bring together the different worlds of interests:

> The highly placed official speaks in the name of developing the French infrastructure and supports the project of the transportation minister – who speaks in the name of government, which speaks in the name of voters. The transportation minister supports Matra's project, and Matra speaks in the name of captive drivers, who support the project of the engineer, who speaks in the name of cutting edge technology. It is because these people translate all the divergent interests of their constituents, and because they meet together nevertheless, that the Aramis project can gain enough certainty, enough confidence, enough enthusiasm to be transformed from paper to prototype. (Latour 1996: 42–3)

Latour's case study focuses on a large-scale, and expensive, public engineering project. However, the small-scale also admits of actor-network analysis. John Law's early actor-network study of a single physics research laboratory (Law 1994) identified the localized networks surrounding individuals and small groups of researchers involved in specific research projects and, in contrast to those studied by Latour, not responsible for making tangible 'products' such as trains. Law's theoretical and conceptual framework has proved influential on subsequent studies as he reconfigures essentialist notions of what people are. Rather than maintaining a version of people as single and bounded entities, Law notes that: 'People are networks. We are all artful arrangements of bits and pieces. . . . We are composed of, or *constituted* by our props, visible, invisible, present and past' (Law 1994: 33). In addition, Law notes that ANT's basis in a post-structural perspective means that social analysis is a process of telling stories, and he foregrounds the contingent nature of the narratives we tell about the world: '[ANT] tends to tell *stories*, stories that have to do with the processes of ordering that generate effects such as technologies, stories about how actor-networks elaborate themselves, and stories which erode the analytical status of the distinction between the macro and micro-social' (Law 1994: 18).

But there is a problem here. 'Network' is a metaphor; networks are arrangements that are socially constructed but also require an external observer to name them as such; they aren't there in the sense that a network of power cables is there. People *aren't* networks, or are only networks if we use a form of nominalism that allows us to redescribe and redefine according to rules that are chosen arbitrarily. Further, the use of spatial metaphors such as 'network' or 'node' provides us with not just a way of ordering objects that we encounter in our research, but a way of interpreting the world; spatial metaphors have an effect on our perceptions. Writing in 1982, before ANT had become established, sociologist Wilhelm Baldamus noted:

There is little doubt that the often remarkable longevity particularly of spatial metaphors is a contributing factor to the survival of obsolete theoretical conventions. The concept of 'network' is of special interest here because it shows that even a metaphor with hardly any explanatory power to start with can maintain its popularity for long periods for no tangible reason. (Baldamus 2010a: 107)

In ANT, Latour has recently attempted to replace 'network', or rather reconfigure it. But rather than the concretization of the metaphor being the problem, it is the attendant ambiguity that he dislikes: 'The word network is so ambiguous that we should have abandoned it long ago' (Latour 2005: 129). This is not the same as the point made by Baldamus – that our definitions of 'network' are invariably synonymous ones, and that, lacking definition, 'network' should be abandoned. On the contrary, Latour does want us to keep the word, but to use it in a different way. Rather than looking for networks in the world, he now proposes that we use 'network' as a tool that helps to describe something, and as a measure of energy and movement:

So, network is an expression to check how much energy, movement, and specificity our own reports are able to capture. Network is a concept, not a thing out there. It is a tool to help describe something, not what is being described. It has the same relationship with the topic at hand as a perspective grid to a traditional single point perspective painting: drawn first, the lines might allow one to project a three-dimensional object onto a flat piece of linen; but they are not *what* is to be painted, only what has allowed the painter to give the impression of depth before they are erased. In the same way, a network is not what is represented in the text, but what readies the text to take the relay of actors as mediators. The consequence is that you can provide an actor-network account of topics which have in no way the shape of a network – a symphony, a piece of legislation, a rock from the moon, an engraving. Conversely, you may well write about technical networks – television, emails, satellites, salesforce – without at any point providing an actor-network account. (Latour 2005: 131)

This rather prompts the question: why **network** and not some other metaphor? And if it is simply the imagery of the net that ANT wants to retain, why does Latour confuse this simple image by dropping so many other metaphors and allusions into his work; social fluid, meshes, circuitry (Erickson 2012)?

ANT cannot be faulted for its identification of the complexity of relationships surrounding knowledges and technologies, and many ANT studies reveal aspects of the generation of scientific knowledge that had been previously hidden. However, the trajectory of ANT has been such that it has moved away from considering the relationship between formal scientific knowledge and the people who produce it, to a focus on more processual aspects of the emergence of specific technologies. There is, of course, nothing wrong with this per se: STS is an area of study that is concerned with *technology*. Yet there is an issue here: formal science and knowledge that emerges from formal science locations that contributes to the construction of networks including technologies is objectified in a way that may not be entirely helpful. In terms of method and fieldwork, ANT encourages a form of mimesis – a re-inscribing and re-labelling of things in the world, but not a mode of understanding. Re-labelling a conversation as a translation between mediators in a network, or a move to achieve control in a struggle between two identities, does not clarify *why* things are happening. And whilst ANT may show complexities of networks, it often does not challenge the power differentials and imbalances in such networks, nor does it consider science to be anything other than another object in a network.

Some STS writers have noted that there are considerations external to specific networks. David Hess, for example, notes that:

> Actors come to networks within cultures that provide them with biases about appropriate forms of knowledge, methodology and machinery. Thus, although an actor-network analysis brings to STS the helpful corrective that shows one way in which structural change is possible, discussions of actor-networks need to be framed by an analysis of culture and power. (Hess 1995: 53)

The stories ANT tells are plausible and illuminating, but they may not be telling us everything we need to know. They can be prone to becoming bogged down in the minutiae of local networks, and this localized focus can lead to an exclusion of analysis of external factors, contexts, cultures and power. In addition, by putting 'science' back into a specific location that can be analysed through the identification of networks that pass knowledge around, STS and ANT may actually be reinforcing the standard account of science, where science is bounded, separated and superior. Often STS remains a form of localized analysis, a new way of telling stories about technologies in specific places and times. Actor-network studies are prone to a number of criticisms and challenges, not least of which is their need to adopt realist positions, as the non-human agents they encounter form part of the same network as humans (Sismondo 2004: 73). A further problem for ANT, when analysed from a cultural studies perspective, is that its focus on discrete networks systematically excludes from consideration exactly the same people that have been systematically excluded from consideration by proponents of the standard account. In other words, ANT suffers from a problem similar to the SSK version of social constructionism: a tendency to reproduce, albeit in different words, the standard, essentialist understanding of what science is and where it resides.

Thought communities and the social production of scientific knowledge

Fleck, working as a microbiologist in the 1930s, produced a remarkable and prescient book that amalgamates historical and social analysis to investigate how scientific communities come to agree on what counts as knowledge (Fleck [1935] 1979). He presents a history of syphilis and scientific knowledge of this disease. He looks at the first appearance of syphilis in Europe in the fifteenth century, how medics at the time made sense of it, and the emergence in the late nineteenth and early twentieth centuries of scientific discourses of syphilis. In particular, Fleck focuses on the Wassermann reaction, a laboratory procedure commonly used as a proof of the existence of the syphilis bacillus, thus a proof of disease. Interestingly, it was not known why the Wassermann reaction worked, yet it became the standard diagnostic tool in the early 1900s and was in extensive use throughout the first half of the twentieth century.

Fleck starts with the point that sociologists of his day were wrong to exempt scientific *knowledge* from sociological investigation. He argues that we should not uncritically accept accumulated progress in scientific knowledge 'as if our way of thought represented an improvement upon the thought styles of previous generations' (Trenn and Merton 1979: 155). He categorically rejects the idea that currently recognized 'facts' are more true, and proposes that we understand 'facts' as being contingent on place and use in a particular location – what he calls a 'thought community' (*Denkkollectiv* – see box 1.5 on p. 21). A thought community is a group of individuals who share ideas, concepts and theories: they share a particular 'thought style'. All scientific communities are thought communities, as are a great many other forms of collective being (families, political party meetings, groups of friends, etc.), so all of us are members of a number of thought communities. Whilst this concept may not have been new, it enabled Fleck to challenge the idea of the 'lone genius' bringing about a scientific revolution on their own. Scientific revolutions take place, but these are collectively generated. Not only that, but also they are premised on social and cultural conventions – i.e. factors external to science – that are shared by members of a community. As Fleck says:

> Truth ... is always, or almost always, completely determined within a thought style. One can never say that the same thought is true for A and false for B. If A and B belong to the same thought collective, the thought

will be either true or false for both. But if they belong to different thought
collectives, it will just *not* be *the same* thought! It must either be unclear
to, or be understood differently by, one of them. (Fleck [1935] 1979: 100)

Scientific knowledge is socially and culturally conditioned. Facts vary
with time and culture – they are a product of thought styles. Thought
styles are specific to thought communities: they are the underlying
pattern which structures the production of knowledge in, for example,
a scientific community.

Fleck's term 'thought style' (*Denkstil*) is central to how and why
facts become validated inside a scientific community; as we think in
different ways we will discover new facts. 'Both thinking and facts are
changeable, if only because changes in thinking manifest themselves in
changed facts. Conversely, fundamentally new facts can be discovered
only through new thinking' (Fleck [1935] 1979: 50–1). There is no implication of 'progress' here. Fleck is deploying a version of nominalism:
the thought style of a thought community is collectively held and is the
basis upon which facts are identified and then validated. Fleck notes
that the modern understanding of syphilis as being caused by a specific
bacterium and being testable by only **one** method means that modern
medical science ignores a myriad of other important things.

> The development of the concept of syphilis as a specific disease is thus
> incomplete in principle, involved as it is in subsequent discoveries and
> new features of pathology, microbiology, and epidemiology. In the
> course of time, the character of the concept has changed from the mysti-
> cal, through the empirical and generally pathogenetical, to the mainly
> etiological. This transformation has generated a rich fund of fresh detail,
> and many details of the original theory were lost in the process. So we
> are currently learning and teaching very little, if anything at all, about the
> dependence of syphilis upon climate, season, or the general constitution
> of the patient. Yet earlier writings contain many such observations. As
> the concept of syphilis changed, however, new problems arose and new
> fields of knowledge were established, so that nothing here was really
> completed. (Fleck [1935] 1979: 19)

Fleck's work anticipated that of a number of important writers. He
was certainly aware of the Vienna Circle style of thought: he explicitly
criticizes it, and his relativistic understanding of scientific knowledge
is similar to the later philosophy of Ludwig Wittgenstein. Indeed, both
men were living in Vienna in 1927. Although there is no evidence to
suggest that they met, if they did, it might begin to explain Fleck's
relativistic understanding of issues such as 'truth' and how different groups of people can have different understandings of the same
thing. Wittgenstein introduces the concept of 'form of life' to describe
the environment in which a particular set of linguistic and gram-

matical usages will come to have meaning, and suggests that there is a range of forms of life available. Fleck's concept of 'thought communities' is similar to this. Fleck asserted that there will be pre-ideas inside a thought community that will prefigure a later, full-blown, idea, and that the adoption of the idea will be through its collectivization in a thought community and the adoption will look like a revolution.

Three main connections can be identified between Fleck's work and more recent social thought. Firstly, his analysis of a discontinuous and non-progressive history of medicine is similar to Michel Foucault's (e.g. *Birth of the Clinic*, or *Madness and Civilization* (Foucault 1973, 1967)). Historical analysis in the Foucauldian style suggests that we simply have different rather than better ways of understanding medical problems, different names and different preoccupations. Foucault, of course, wants to point to these different perspectives being socially and culturally determined.

Secondly, the connections to Thomas Kuhn's work are quite clear: the idea of an underlying paradigm/thought style that directs the activities of formal scientists is common to both writers (Kuhn's history of science and his conception of paradigm are discussed in more detail in chapter 4). Indeed, Kuhn did make a very brief mention of Fleck's work in the preface to *Structure* (Kuhn [1962] 1970: vi), noting that he had chanced on an unknown German monograph of 1935 that 'anticipates many of my own ideas' and which he is 'indebted to . . . in more ways than I can now reconstruct or evaluate' (ibid.: vii). This is, perhaps, disingenuous in that the main ideas of Fleck's work – particularly the idea of incommensurability – feature heavily in Kuhn's *Structure*; the exclusion of any mention of Fleck in the whole book is, in Baldamus's words, 'difficult to understand' and prompts us to feel a 'sense of injustice' (Baldamus 1976: 43). However, the assertion of Fleck anticipating Kuhn's ideas hides something quite important: Kuhn's work is primarily based on the history and philosophy of science; Fleck's is a *sociological* account of how scientists actually do their work and construct their theories through social processes. As well as there being no mention of Fleck in *Structure* there is no sociology or social analysis in *Structure*. Kuhn's work is much less radical than Fleck's, in that it describes the scientific community as being isolated and external to society, following the standard account of science and society, whereas Fleck shows through a large amount of evidence the connections between different exoteric and esoteric thought communities that bring the social and society into the heart of the production of scientific knowledge. Where Kuhn can tell us what has happened, Fleck can tell us *why* it has happened and, perhaps most importantly, why it is that certain forms of scientific knowledge emerge at certain times. We can look at the organization structure of a scientific community, but this does not tell us where a paradigm may come from; Fleck emphasizes that the individual scientist cannot escape the social constraints (*sozialer Denkzwang*) on all thought processes. So where we can identify, on the surface, individual scientists making suppositions, or speculating about what experiments to carry out, arguing about objectives and so on, we need also to consider what thought styles, discourses, experiences and conditions both internal and external to the scientist's community precede these actions. This has wider implications: Kuhn's theory does not promote openness in science, as it demarcates and sets a boundary around the scientific community and insists that this is necessary for its functioning (Fuller 2003). In contrast, it is Fleck's work that shows that openness is vital, and that we need more openness if we are to understand science in our society.

Finally, Fleck's work anticipated the current trend in science studies, that of social constructionism. Fleck argues that scientists are involved in 'making up' knowledge. We would possibly identify Fleck as being at the soft end of social constructionism in that he does think that facts are real, but recognizes that they are culturally validated by a thought community.

Conclusion

The standard account is significant because, from a sociological perspective, it does describe many aspects of the meanings attached to actions and the motivations for action that formal scientists have. This presents us with a major problem. As I have said, STS feels that the standard account is wholly inadequate for describing what is happening in scientific knowledge production, and as we will see there are aspects of the standard account – particularly when connected to realism – that seem to be flawed or contradictory. But from a sociological perspective we need to respect the fact that the explanations for action that people express are, at the very least, meaningful to them and may be causal factors in further action. If the meanings that are attached to actions by scientists do not match up with those that STS and sociology of science researchers consider to be valid, what should we do? Do we simply reject scientists' meanings as being a kind of 'false consciousness'? Can we confidently say that scientists are wrong in their understanding of what they are doing in their work? It is unlikely that we would do this to other groups of people in contemporary society. Can this paradoxical situation be explained?

Scientism in a technoscientific society becomes a dominant mode of thought and it has significant, if differential, effects across communities and institutions. In a scientific community scientism has an even greater resonance and acts to reinforce tradition and repetition of linguistic constructions that 'explain' the world. Most of us, formal scientists included, don't think in great depth about what we are doing, what the world is made of, how we know about the world; we just get on and do it in the ways that we always have. Scientism, the standard account and realism have an elective affinity that is enhanced and perpetuated by the structure of the scientific community as a whole, and, particularly, esoteric thought communities where to be a member you have to conform to a very precise way of thinking, as well as holding the general tenets of the standard account. The elective affinity between these things has, it would appear, a kind of compulsion that requires all aspects to be adopted, although perhaps not uncritically, for membership of the thought community.

In contrast, social science operates in a slightly different way. Whilst it is obvious that esoteric thought communities in the social sciences require members to share a thought style, and equally clear that the exoteric impinges directly on the esoteric (just think of the battles that feminists had to wage inside the social sciences to get women's voices heard), the social sciences' lack of a shared paradigm or general thought style sets them apart from the formal sciences. Rather than finding elective affinities between an account of what social science does/is trying to do, what the best way of understanding the phenomena we encounter is and which theory we will use to identify those objects, we find much looser and vaguer versions of these things unevenly distributed across a disparate collection of esoteric and exoteric thought communities that are loosely grouped together as 'the social sciences'. But here arises yet another paradox. Fleck identified the way that the formal sciences grew, at least in terms of the amount of knowledge they produced, as being through error:

> The following facts are therefore firmly established and can be regarded as a paradigm [*Paradigma*] of many discoveries. *From false assumptions and irreproducible initial experiments an important discovery has resulted after many errors and detours.* The principal actors in the drama cannot tell us how it happened, for they rationalize and idealize the development. (Fleck [1935] 1979: 76)

The formal sciences thus proceed through collective error and subsequent correction and rationalization. In contrast, the social sciences proceed through a *lack* of error, individual and collective, in that social scientists have no way of knowing when they are wrong except in the most blatant of circumstances (Baldamus 1976). They do, however, employ plenty of rationalization.

SSK, STS and ANT have common roots in a relativist project. Despite these shared roots, they have taken a path that serves to reinforce dominant accounts of science and the discourse of scientism;

as they challenge the standard account they have been instrumental in strengthening the idea that science is separate and has an essence. In contrast, Wittgenstein's thought tells us that we need to have a much more complex and considered understanding of what science in society means, and Fleck's work suggests that to understand science fully we should look at scientific communities and the relationship between internal, esoteric knowledge and external, exoteric knowledge. As well as considering these relationships we also have to consider the history of science, the back story that constructs a significant part of the picture of science visible today. It is to this that we will now turn.

Further reading

Introductions to the philosophy of science

All of the following provide good overviews of the main themes of philosophical analysis of science:

Bortolotti, L. (2008) *An introduction to the philosophy of science*, Cambridge: Polity.

Chalmers, A.F. (2013) *What is this thing called science?*, Buckingham: Open University Press.

Couvalis, G. (1997) *The philosophy of science: science and objectivity*, London: Sage.

Hacking, I. (1983) *Representing and intervening: introductory topics in the philosophy of natural science*, Cambridge: Cambridge University Press.

Social constructionism

Social constructionism is a huge topic, and it extends much further than analysis of science and technology. The following remains the best starting point and provides in-depth analysis of how sociology of science deploys social constructionism:

Hacking, I. (1999) *The social construction of what?*, Cambridge, MA: Harvard University Press.

Actor-network theory

Bruno Latour set out ANT in his 1987 book:

Latour, B. (1987) *Science in action*, Cambridge, MA: Harvard University Press.

More recently he has set out a more complex theoretical framework, and provided a strong critical appraisal of ANT:

Latour, B. (2005) *Reassembling the social: an introduction to actor-network-theory*, Oxford: Oxford University Press.

Ludwik Fleck

The best starting point is the introductory essay in the translation of Fleck's 1935 book *Genesis and Development of a Scientific Fact*:
Fleck, L. (1979) *Genesis and development of a scientific fact*, Chicago: University of Chicago Press.

There is also a major collection of essays on Fleck:
Cohen, R.S. and Schnelle, T. (eds) (1986) *Cognition and fact: materials on Ludwik Fleck*, Dordrecht: Reidel.

Fleck's entry into Anglophone sociology and philosophy of science came through the writings of Wilhelm Baldamus, who provided the first English-language translations of Fleck's work in:
Baldamus, W. (1976) *The structure of sociological inference*, London: Martin Robertson.

Baldamus trenchantly analysed Thomas Kuhn's uptake and use of Fleck's work in a number of publications:
Baldamus, W. (1977) 'Ludwig Fleck and the development of the sociology of science', in Gleichmann, P.R., Goudsblom, J. and Korte, H. (eds) *Human figurations: essays for Norbert Elias*, Amsterdam: Stichting Amsterdams Sociologisch Tijdschrift. pp. 135–56.
Baldamus, W. (1979) 'Das exoterische Paradox der Wissenschaftsforschung: ein Beitrag zur Wissenschaftstheorie Ludwik Flecks', *Zeitschrift für allgemeine Wissenschaftstheorie*, 10, 2, 213–33. [Trans. and repr. in Erickson, M. and Turner, C. (eds) (2010) *The sociology of Wilhelm Baldamus: paradox and inference*. Farnham: Ashgate.]

Ludwig Wittgenstein

Wittgenstein's philosophy used in this book comes from the later period of his work. The most important text is:
Wittgenstein, L. [1953] (1958) *Philosophical investigations*, Oxford: Blackwell.

Wittgenstein's work can be hard to interpret, and good secondary texts can be helpful:
Hacker, P.M.S. (1997) *Wittgenstein*, London: Phoenix.
Monk, R. (2005) *How to read Wittgenstein*, London: Granta.

Both of these books provide quick and lucid introductions to how to approach Wittgenstein's work. For more depth and detail:
Hacker, P.M.S. (1986) *Insight and illusion: themes in the philosophy of Wittgenstein*, Oxford: Oxford University Press.

If you are interested in Wittgenstein's remarkable life, Ray Monk's biography is the most comprehensive and readable account:
Monk, R. (1991) *Ludwig Wittgenstein: the duty of genius*, London: Vintage.

4

History

And whereas sense and memory are but knowledge of fact, which is a thing past and irrevocable; *Science* is the knowledge of consequences, and dependence of one fact upon another; by which, out of that we can presently do, we know how to do something else when we will, or the like another time; because when we see how any thing comes about, upon what causes, and by what manner; when the like causes come into our power, we see how to make it produce the like effects. . . . But yet they that have not *science*, are in better, and nobler condition, with their natural prudence, than men, that by mis-reasoning, or by trusting them that reason wrong, fall upon false and absurd general rules. . . . To conclude, the light of human minds is perspicuous words, but by exact definitions first snuffed, and purged from ambiguity; *reason* is the *pace*; increase of *science*, the *way*; and the benefit of mankind, the *end*.

Thomas Hobbes, *Leviathan* ([1651] 1968)

Thomas Hobbes's 1651 book *Leviathan* sets out his understanding of society, how it is possible, and what constitutes the human beings that populate it. It is a powerful expression of a new kind of universalism – secular humanism: that the world is knowable to people, not just to god(s), a dramatic departure from the theological universalism that characterized European thought to this point. It is also, perhaps, the first systematic sociology text and the first attempt at a social psychological understanding of the human condition. Being the first doesn't necessarily make you the best, but in many cases, and particularly that of Hobbes, it often means that you get to choose the terrain and the objects of inquiry for a discipline. If for no other reason, Hobbes's work is worthy of study because of this. My concern here is with how Hobbes formulated his philosophy in his seventeenth-century world, the constituent elements of the new universalism, the new way of making sense of everything in the world. Hobbes was inspired by, and deployed in his philosophy, an understanding of science that emerged at a crucial time in Western European history. Hobbes put science, scientific knowledge and scientific method at the heart of his systematic analysis of society and people, and his heirs have followed this trajectory ever since.

Historians of science refer to 'the scientific revolution' as being the period in European history when the 'conceptual, methodological and institutional foundations of modern science were first established' (Henry 1997: 1). This account sees the discoveries of the sixteenth century as scene setting for the main phase of the scientific revolution in the seventeenth century, which culminated in the consolidation and founding of modern science in the eighteenth century. Historian of science John Henry notes that 'scientific revolution' is a term of convenience for historians of science, but also argues that it is a real thing in that knowledge of the natural world was very different in 1700 from that in 1500; there is a general consensus amongst historians of science that the sixteenth and seventeenth centuries did see a remarkable change in knowledge, although the degree of continuity and the significance of the change are still hotly debated (Principe 2011). Hobbes's position in this change in knowledge is interesting. He is best known for his work in political theory, but *Leviathan* – his magnum opus – is much more than just a description of how politics should be conducted; it is a systematic and comprehensive analysis of the self, society and nature. At the heart of *Leviathan* Hobbes has a clear idea of how we should understand the world: as matter in motion (*Leviathan* ch. 2). That he should be interested in science and physics in particular is no surprise; he worked as amanuensis to Francis Bacon between 1621 and 1629 and must have been familiar with Bacon's theories of experimentation and induction. However, as C.B. Macpherson succinctly puts it, 'Hobbes's conversion to science, when it came in 1629, was to a different concept of science' (Macpherson 1968: 17), a version of science deeply informed by geometry and the work of Galileo.

John Aubrey, Hobbes's first biographer, claims that Hobbes swore ('By G–' said he, 'this is impossible!' (Aubrey 1982: 150)) when he first realized, at the age of forty, the immensity of the possibilities that geometrical analysis presented. Being able to demonstrate the truth of a complex proposition by reference back to simple axioms 'made him in love with Geometry' (ibid.). Hobbes became obsessed with geometry and motion, so much so that he went on a 'pilgrimage' to Florence to meet Galileo during an extended continental trip (1634–7) (Macpherson 1968: 18). This was undoubtedly a seminal moment in the development of Hobbes's ideas and for the development of social science as a whole. However, Hobbes's personal discovery of geometry antedated his meetings with Galileo by some five years, so there may be additional factors in Hobbes's intellectual transformation. D.D. Raphael argues that the idea of matter in motion having applicability other than to celestial bodies may have come from Hobbes's reading of the work of John Harvey, who discovered the circulation of the blood. Harvey wrote 'I began to think whether there might not be a motion as it were in a circle' (Raphael 1977: 22). Harvey is applying a geometrical principle in his physiological research; Hobbes's philosophy correlates physics with physiology and extends this move to analysing human action and human mind (Raphael 1977: 23). But Hobbes was not distinguishing this form of knowledge as being separate or only accessible to a select group; science, and scientific knowledge, were something that *all* could aspire to and participate in. Equally, science was not something that need only apply to the physical world: we could understand society and the human mind and consciousness using this form of knowledge. Hobbes's philosophy is reductionist in form, bringing everything, ultimately, to the level of matter in motion. Even our consciousness, for Hobbes, is not real at all, but an appearance that is a response to our bodies in motion. It is not surprising that, with roots stretching back to this significant meeting of ideas, social science has such a persistent relationship with the natural sciences, and I would argue that the pernicious scientism that infects contemporary society has its origins at this point. From this moment on, science becomes a tool that philosophy and subsequently social science could deploy to make sense of the world; science becomes the form of knowledge that philosophy and social science, and other forms of knowledge, defer to in matters of truth. At the time Hobbes was writing there was no separation of arts and humanities from the natural sciences. The secular humanistic knowledge that he articulated was unitary and universal – an expression of humanistic universalism in contrast to the theological universalism it was rapidly replacing.

Hobbes's definition of science is down to earth and a reasonable reflection of what many people today consider science to be; I think he would have agreed with the standard account of science described in chapter 3. What *is* different is where he located science and who he thinks is deploying science: it *isn't* just scientists (or, given that the word 'scientists' had not entered the English language at that time, people we could call 'natural philosophers' or some other intellectual elite); it is, potentially, *everyone*. For Hobbes, science is something that is demotic, a way of doing knowledge that all people of reason can do. How did we get from that situation, of science as something that we can all, if we put our minds to it, do, to the current situation of science (seen as) separate from society and the province of a small group of highly specialized people? How did we get from science, as Hobbes conceived it, to formal science?

It is an interesting question: how is it that we have separated thought and knowledge into separate categories, where did these categories come from, and how have these categories shifted and changed over time? Too much for the scope of this book, certainly. But the institutional analysis of what we have done with knowledge through modernity is important: taking a sociology of knowledge approach will help us to understand how the history of science is not a tale of a seamless progression of knowledge, building block upon block with the best scientists of any day 'standing on the shoulders of giants'. Rather, we will find that when we look from the outside at the history of science, we will see disciplinary differentiation and consolidation taking place at the same time as external factors push knowledge to be formalized and institutionalized such that certain imperatives of power come to be promoted. And when we look internally at the history of science, we find a fragmented and discontinuous narrative that is selectively sampled to provide each generation of scientists with a back story that provides a foundation and justification for their current activities. Finally, when we look at the majority of histories of science, be they specific or general, for a 'professional' or a 'lay' audience, we see a story of the emergence of a form of knowledge and a group of people that are superior, separate and selfless.

One thing that has clearly happened in the 400 intervening years is that the word 'science' has changed its use; the rules for deploying the word have changed, the language-game of science has changed. But this happens inside a form of life that itself is changing, and these changes are not occurring in isolation due to some inner logic or 'natural' process. Our complex form of life changes in response to many, many influences. We need to look at this complex of influences impinging on us and our knowledge across time to understand how we have ended up with a particular version of science now, and a particular hierarchy of knowledge in our culture.

We need to be wary that we don't simply see these shifts and changes as occurring in isolation; the 'scientific revolution', the rise of scientific modes of thinking, the idea that science is a different and better mode of understanding all emerge in a context that we need to understand. It is the context surrounding thought and knowledge that is significant in determining its trajectory. This is the clear message that comes from the work of Immanuel Wallerstein – not a traditional historian of science, but a historian who tries to see change in much wider than usual historical perspectives. Wallerstein's work in world-systems theory covers, as the name implies, a huge number of topics and crosses disciplines, and a full treatment of his theory is not possible here. In the following section we will focus on Wallerstein's account of the emergence of scientific universalism and the bifurcation of knowledge. Where does the split of 'science' from 'the rest' come from, and why do we have it? We'll use his theory to reconstruct an alternative history of science, one that puts science into a wider social, cultural and economic context.

Wallerstein's history starts not with the past but with the present: what do we need to know? We need to understand the world we live in, but we are constantly hampered in this because we use a range of different disciplines to look at the world. To us it may seem sensible to use psychology to look at the mind and economics to look at the economy, but for Wallerstein as we partition the world into boxes we don't see that these boxes are 'constructs more of our imagination than of reality. The phenomena dealt with in these separate boxes are so closely intermeshed that each presumes the other, each affects the other, each is incomprehensible without taking into account the other boxes' (Wallerstein 2004: x). These separate boxes – the academic disciplines – are an obstacle to understanding, but are also generated by the operation of the capitalist world-system.

We tell ourselves a mythical story of universities, that they have a tradition going back to medieval times, and reinforce this with ceremonies (congregations) and regalia (gowns and mortar boards). But it really is a mythic history. The university system we are familiar with was a creation of the late nineteenth century; it is a product of modernity and unlike a medieval university, which was a site for the validation, dissemination and legitimation of theological knowledge, is characterized by a full-time staff, centralized bureaucracy, formalized degree-awarding system and, crucially for Wallerstein, a compart-mentalized structure, with different disciplines occupying different departments and faculties.

The medieval university was divided into four faculties: medicine, theology, law and philosophy (Wallerstein 2004: 3). In the later part of the nineteenth century the modern university's faculty of philosophy split into two: the 'sciences' and the 'humanities' or 'arts' (the shared root of degree programmes in a putative faculty of philosophy is the reason that all higher degrees at UK universities are styled 'doctor of philosophy' (PhD)). The modern university is a secular humanist response to the theological universalism of the late medieval period, but by the late nineteenth century universities were in the grip of scientific universalism. This is unsurprising, given that by this point in industrial societies scientists were able to point very clearly to the contributions they had made to capital, industry and general progress. Science was the code word for achieving progress, a claim that is unremarkable to us now but was in the nineteenth century a marker of a shift in the value-systems that dominate knowledge (Wallerstein 2004: 74). Scientists could, and did, make claims for social prestige, and for financial support for their institutions, based on their claim that scientific knowledge can be transformed into technology and thereby contribute to progress. This restructuring and refinancing of universi-ties was a struggle between competing academics – a social process in essence. But the stakes were high; after all, universities are social institutions that legitimate certain types of knowledge and particular forms of education throughout a society's education system, including schools. This was a struggle for who gets to decide what is 'true', and who gets to socialize our young people and with what forms of knowledge (Wallerstein 2006: 63). A settlement emerged in the modern university where the sciences would be given the role of finding legitimate truths. The humanities and arts would be given the role of deciding on what is good and/or beautiful. The social sciences – newly emerging at this time – would sit uneasily between these two. The split of scientific from humanistic knowledge and disciplines was stark and has had widespread consequences. 'Never before in the history of the world had there been a sharp division between the search for the true and the search for the good and the beautiful. Now it was inscribed in the structures of knowledge and the world university system' (Wallerstein 2006: 63). This is the origins of the 'two cultures' debate, made famous by C.P. Snow in the 1950s (Snow 1959; Snow and Collini 1993): the idea that there is a separation of knowledge cultures into arts and sciences that cannot be bridged, and that there are thus two separate (and ranked) ways of understanding the world. We may need to remind ourselves quickly: these changes in knowledge structures are not a result of internal evolution or gradual accretion; they emerge from a process of intellectual struggle whereby the forms that have been most useful to the maintenance and development of the modern world-system come to the fore and flourish (Wallerstein 2006: 58ff).

The dominance of scientific universalism – the idea that scientific methods provide the only means of accessing truths in all cases, and scientific knowledge is a universally applicable form of knowledge that is the sole repository of legitimate truths – has had significant consequences. Not only does it contribute to and in part constitute the dominant worldview of scientism that is abroad in Western industrial societies, but it also legitimates structures of power, dominance and discrimination, according to Wallerstein. 'Scientism has been the most subtle mode of ideological justification of the powerful' (Wallerstein 2006: 77). It presents scientific universalism as ideologically neutral, as unconnected to culture, as being justified by the power of good it can do for society. It is hard to argue against it, particularly as we spend so much time arguing for it (e.g. begging politicians to listen to climate change scientists). But it is this alleged neutrality that can be, according to Wallerstein, damaging to society.

As science assumed its pre-eminent position, and as it established the theoretical virtue of meritocracy (position being awarded purely on the basis of competence which could be measured by objective criteria), the practitioners of science were elevated to a high moral status. It looked as if those in positions of power in the institutions of science were morally entitled to be there. And those inside these social institutions become autonomous judges of their own value and recruit-ment. In wider society the idea that access to all social positions was achieved by merit becomes widespread currency – despite the evidence to the contrary. We move to a position where we think that those who are not in high social positions only have themselves to blame: inher-ent incompetence, cultural provincialism, perverse will (they simply will not help themselves!). Not only that, we begin to see the world in this way: 'backward' nations only have themselves to blame, and so on (Wallerstein 2006: 76–9). That this can be socially corrosive is obvious. But there is another, further twist: if we separate the 'good' from the 'true', it allows those involved in quests for the truth to ignore value considerations whilst re-inscribing and re-imposing their own value-system from inside their own thought communities.

There is a further consequence of this, for the history and histori-ography of science. To reflect its current status and its role as locator and arbiter of the truth, formal science must represent its history as a seamless and value-neutral progression towards the present and into the future. This is visible in almost all histories of science, regardless of the audience they are directed at. They present a standard account of the history of science, and there are strong connections to the standard account of scientific knowledge and scientific method that we have already encountered. The standard account of history, like the other accounts, has begun to be challenged, contested and reworked in recent years.

The standard account of the history of science

The standard account of the history of science is a story of progress, accu-mulation of knowledge and increasing discovery. It is a triumphal tale which starts with human ignorance of the workings of the natural world and presumably ends with total human understanding of nature. This history of discovery and invention is constructed with the benefit of hindsight, and obscures aspects of scientific history that do not conform to the ideal being promoted (for example, phrenology – a cornerstone of nineteenth-century sciences of the mind, but now wholly discredited – is usually omitted from contemporary histories of science). Such a history gives a generous role to 'scientific thinking', seeing it as the driving force behind all technical and technological development, and as a motor for much human social development. It implicitly defines science as being any human endeavours that have attempted to under-stand and control nature. Moreover, it frequently reinforces the myth of the creative power of the lone scientific genius.

Such an approach is taken by Isaac Asimov in his *Asimov's Chronology of Science and Discovery* (Asimov 1989). From 4,000,000 BC to the late

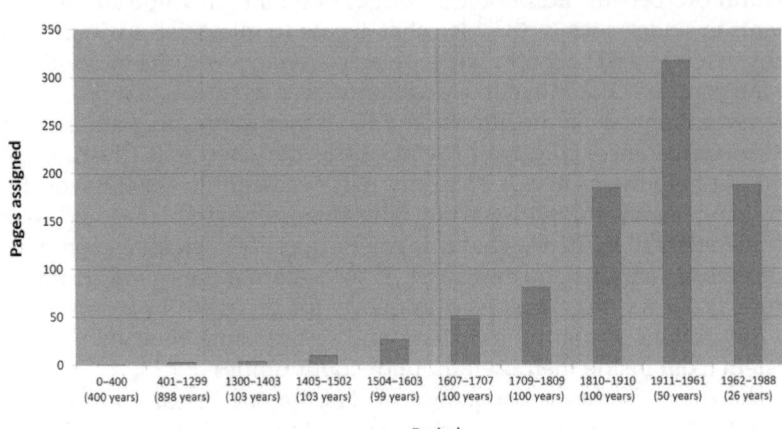

Figure 4.1 Asimov's chronology of science and discovery. Note – dates BC have been omitted (Asimov covers 4,000,000 BC to AD 0 in sixty pages).

1980s, Asimov presents a continuous thread of discovery and invention that begins with stone tools, passes through fire, agriculture and the windmill, continues with the modern age discoveries of the motion of the Sun (1783), pepsin (1836), psychoanalysis (1893), the Salk polio vaccine (1954), and ends with the discovery of the greenhouse effect (1988). The notable feature of Asimov's account is the increase of pace that we can see in terms of specific discoveries being made. By his account, not much happened in terms of scientific discovery between 4,000,000 BC and about AD 1600. However, after that point, science really speeds up in terms of discovery and invention. We can display this graphically by counting how many pages each slice of history is assigned and placing these on a histogram (figure 4.1).

Within this seamless narrative there is no pause to comment on the social causes and contexts of the emergence of these discoveries and achievements. Asimov's book is typical in its tone of many accounts of the history of science. For example, Brian L. Silver's *The Ascent of Science* (Silver 1998) gives away its orientation towards science in its title. The contents provide a story of the increase in rationalization and knowledge of the natural world – again, a standard version of the triumphal progress of science. Silver focuses on the role of lone scientific geniuses and their discoveries, reinforcing a popular stereotype which is, arguably, a significant distortion of scientific experimental work. Like Asimov's, Silver's account provides much detail of specific inventions and theories. However, the representation of science as a remorseless accumulation of knowledge made by individuals working alone serves to reinforce the standard account of science as superior to other forms of knowledge, and reinforce the idea that scientific knowledge provides a solid and, almost, complete account of the workings of the natural world. The history of science as triumphal progress shows the gradual, and then increasingly rapid, completion of the inventory of humankind's knowledge of the natural world. Such a history both inspires confidence and reassures, suggesting that if no specific knowledge of natural phenomena exists at the moment, the scientific method will be able to provide such knowledge through extension of experiment and theory in the future.

John Gribbin's *Science: A History 1543–2001* (Gribbin 2003) is a history of science from the Renaissance to the time the book was written, based on a biographical approach. Most individual chapters and subsections in the book are based on the formula '*X* discovers/invents *Y*'. Not surprisingly the major figures from the pantheon of the standard history of science appear: Galileo, Newton, Faraday, Einstein and so on. This is an account that tells the story of the history of science in a predictable and triumphal way. Scientists line up in a row and extend each other's discoveries about the natural world. So: Copernicus, Galileo, Kepler and so on. It is worth noting that the book is rather slapdash, with a number of errors of fact, such as saying the inventor of logarithms was English (Napier was Scottish and a university in Edinburgh is named after him) or claiming that 'plants are mainly made of carbon dioxide' (ibid.: 220 n.1), and with some rather large omissions, particularly from disciplines Gribbin is perhaps not so familiar with, such as biology; the 650-page book makes no mention of Pasteur, Koch, Fleming or Salk. Gribbin's book serves to inform the reader about science, but also reinforces the idea that scientific discoveries and breakthroughs are the result of the work of lone geniuses. The book quite explicitly defends the idea of 'the development of science in essentially incremental, step-by-step terms' (Gribbin 2003: 614).

In the following sections we will focus on a single example which illustrates the standard history of science and more recent challenges to it. Throughout the following account we will see that the history of science and theories of scientific knowledge are closely linked.

A history of sunshine 1

'How does the Sun shine?' appears to be a simple and straightforward question that has a simple and straightforward answer: the Sun shines because a process of nuclear fusion is taking place inside it, releasing vast amounts of energy and thus providing sunshine for our planet. This answer is part of the general societal knowledge of how the world works that is taught in school science classes and receives additional backing from the media. However, if all we can say is 'the Sun works by nuclear fusion', even if we can give a definition of nuclear fusion, we may not have a particularly sophisticated or even 'scientific' understanding of how the Sun works. But again, many would agree that if we really had to we could get more detail, either by looking at science textbooks, or by talking to someone with more knowledge – a physicist or astronomer, for example. Our particular lack of knowledge of the detail of how the Sun works is not indicative of a general lack of knowledge: we know that somewhere, some 'scientists' will know how the Sun works, and will be able to explain it. It is almost inconceivable that 'scientists' do not know how the Sun works, and most people are quite happy to leave it to them to possess the scientific knowledge that explains the fine detail. For most of us it is enough – perhaps even more than enough – to say that 'the Sun works by nuclear fusion'.

But is this correct, and is it a fair assumption? Can we be sure that that is how the Sun works? If we are to answer this question of the correctness (even truthfulness) of nuclear fusion being the cause of the Sun shining, then we need to appraise the elements that make up the statement 'nuclear fusion makes the Sun shine'. This is not easy, not least because nuclear fusion is a complex process.

Box 4.1 Nuclear fusion

Hydrogen and helium are the two main elements in the Sun. Nuclear fusion is the process by which hydrogen atoms fuse together to create helium atoms and release energy. In crude terms four hydrogen nuclei fuse together to produce one helium nucleus. In the total reaction, for each helium nucleus created, two gamma rays, two electrons and two neutrinos are produced (see figure 4.4 on p. 125).

'Overall, fusion in the Sun converts about 600 million tons of hydrogen into 596 million tons of helium every second: the "missing" 4 million tons of matter becomes energy in accordance with Einstein's formula $E = mc^2$. About 2% of this energy is carried off by the neutrinos' (Bennett et al. 1999: 473).

When we begin to examine this knowledge, what is it we are actually appraising? If we say that it is true that the Sun works by a process of nuclear fusion, are we admitting the validity of the theory of nuclear fusion, or of the evidence that was used to compile that original theory, or of the data that physicists have collected when observing the Sun, or of something else – perhaps the status of the person who is telling us the story? (This list could go on and on.) Truth may not be the most appropriate thing to look for. Perhaps rather than asking whether the story of nuclear fusion we are told is true or false, we should ask what makes up the story, what its components are, and how they fit together. Arguments about truth are rarely productive: Wittgenstein tells us that usually our definitions of truth will vary according to our circumstances. Rather than looking to appraise truth or otherwise, it would be better to look at what components are being used to construct statements that we subsequently accept as being true. So, we really need to ask what scientific knowledge is, and specifically, what scientific knowledge of the Sun we have. Then we need to look at how that knowledge emerges from an esoteric scientific community and enters exoteric-oriented science communications, and then our everyday world, in a way that provides us with a very strong and secure story about how the Sun works.

It is necessary to note the parallel here to the examples of laboratory work: a group of workers carry out experiments, turn their results into scientific papers and disseminate this knowledge to their scientific community. This knowledge doesn't simply stay in that location; it emerges in a range of forms and becomes what outside observers see as being formal science, or just 'science'. We are faced with a similar case here: there is a chain that extends from the theory and experiments of astronomers and physicists investigating how the Sun works that extends from localized thought communities to the 'lay' public, via a number of institutions and through a range of different media (e.g. scientific papers, university-level textbooks, school textbooks, reports in popular science journals such as *New Scientist* or *Scientific American*, documentaries on TV, newspaper reports and discussions held in informal, non-scientific settings, to name but a few). Very specific and detailed knowledge produced inside an esoteric thought community is transformed into generally held knowledge in society, a process of exoterization. In addition, all scientific knowledge comes with a 'history': the understanding that scientists have of particular phenomena is contingent to at least some degree upon what they knew in the past, and that is reflected in their shared style of thought. Science often looks as if it is synchronic, as if the knowledge we have of the world has just emerged recently, but there is a strong diachronic component to all scientific knowledge: it has a history, a back story. This situation can be unpacked by looking at an example: theories across time of how the Sun works.

The Sun is the local star to the planet Earth, and the energy that comes from the Sun is responsible for **all** life on Earth. Everything that lives on the Earth derives its energy, ultimately, from the Sun. Not only that, but all the artefacts and objects that we see around us are a product of the energy of the Sun. (Actually, this is not strictly correct. Energy released by radioactive decay is not derived from the Sun. But you get the point – no Sun, no life as we know it. And no DVDs or smartphones either.) For that reason the Sun has been a major object of inquiry for scientists for centuries, and for humankind throughout history. Scientific studies of the Sun have also been an important object of inquiry for sociologists of scientific knowledge. Trevor Pinch's *Confronting Nature: The Sociology of Solar-Neutrino Detection* (1986) is a detailed sociological analysis of the physics of solar-neutrino detection from a social constructionist perspective. A quick overview of the history of scientific investigations into how the Sun works reveals some interesting points.

Ancient formal thought saw the Sun as being a kind of fire – a lump of wood or coal, maybe a glowing rock. Suspended in the heavens, it would be lit in the morning and put out at night, carried across the sky by some supernatural being. This theory explained why the Sun was not visible at night, and why the Sun was warm. This is a pretty good explanation in scientific terms: it is based upon observation and experience and, probably, some hard evidence. Ancient scholars may have observed burning objects falling from the sky – meteorites – and may even have found examples on the ground and studied these. It would be plausible to assume that these pieces of 'fire from heaven' were a part of the Sun: a good empirical deduction, unfortunately leading to the wrong conclusion. From the first written records until the 1800s, not much changed in terms of explaining how the Sun worked. Of course, it was learnt that the Earth orbits the Sun, that it doesn't go out at night, that it is a long way away, and that it is very very big. But gaining this knowledge about the Sun didn't help to explain how it worked. Some process of conventional burning was still assumed to be correct.

In nineteenth-century industrial societies there were two competing theories about how the Sun worked. The first, earlier, thought that it was a big cooling ember that worked in a similar way to burning coal or wood. This was another good theory in the sense of being plausible: it matched experiments that could be carried out on Earth in terms of working out how long it would take for a burning object the size of the Sun to cool down. Experimental data collected from looking at incandescent substances in laboratories could be extrapolated to provide a 'guesstimate' for the age of the Sun. Results from this method were very promising: according to such experimental results the Sun had only been burning for a few thousand years, which matched the current Christian-influenced thinking as to the age of the Earth. The theory of how the Sun worked was validated by experimental evidence that matched another theory that was not directly connected to the first theory – thus providing additional validation. However, it soon became obvious that there were flaws in this theory: chemical burning could not produce the amount of radiation from the observed mass of the Sun, and if the Sun was actually a cooling ember it must have been an awful lot hotter just recently, i.e. within the last few thousand years, too hot for life to survive on Earth. The theory had the advantage over previous ones that as well as using observation, both of the Sun and of meteorites, to collect evidence, it combined these with experiments to measure the rate of burning in the Sun. Its failure came from starting from a point that, whilst plausible, led to impossible conclusions. The theory itself was disproved, rather than the evidence being discredited.

The second, later, theory of the nineteenth century was more robust. It focused on gravitational cooling, whereby the Sun generates energy by contracting in size (formally known as Kelvin–Helmholtz contraction after the scientists who proposed the mechanism). The Sun is a massive body that generates huge gravitational forces which pull material from the outside towards the centre, thus reducing it in size. According to this theory, as the Sun shrinks it loses gravitational potential energy and converts it into thermal energy, which is emitted – hence the heat of the Sun's rays. Because it is so big, the Sun would only need to contract by a little each year to maintain its temperature, and by such a small amount its shrinkage would have been imperceptible to nineteenth-century technology, i.e. unmeasurable. Calculations showed that the Sun could emit radiation from gravitational cooling at its present rate for about 25 million years (Bennett et al. 1999: 467). However, by the mid-nineteenth century geologists already knew that the Earth was much older than that, so the theory had to be wrong (or the geologists had misinterpreted the fossil record; a plausible position to hold, particularly given the controversial implications of a 'very old' Earth with respect to dominant religious thinking at the time). Again, there is a failure of theory, not of evidence or measurement of data. All theories up to this point had used good scientific principles – observation, formulation of hypothesis, attempts to test the hypothesis – to try and validate the theory being tested. All had failed, yet we can see a key principle when analysing these failures: that as science proceeded through time it became more and more unified – discoveries in one area, such as geology, could be used to confirm or reject theories in another area, e.g. astrophysics.

So what is the right theory? The 'correct' answer, the 'true' answer to why the Sun shines was discovered only in the 1920s by the British physicist Eddington, applying Einstein's theory of relativity:

> Einstein's theory included his famous discovery of $E = mc^2$. This equation shows that mass itself contains an enormous amount of potential energy. Calculations immediately showed that the Sun's mass contained more than enough energy to account for billions of years of sunshine, if only there were some way for the Sun to convert the energy of mass into thermal energy. It took a few decades for scientists to work out the details, but by the end of the 1930s we had learned that the Sun converts mass into energy through the process of **nuclear fusion**. (Bennett et al. 2010: 471)

This quotation from a current university-level textbook on astronomy and astrophysics is particularly interesting. Firstly, it is quite triumphal – it says we now know the truth, and implies we were foolish in the past. It also explains the phenomenon of the Sun shining by reference to the paradigm of physics in contemporary scientific thought, i.e. Einstein's theory of relativity. It implies that this problem has been solved and completed, pretty much by Einstein's work in the early part of the twentieth century.

This account looks similar to most other histories of science that describe gradual progress towards finally knowing the truth about phenomena, and being able to piece all of these facts together under a single, overarching theory. It is useful to have a general understanding of this overarching theory, based on Einstein's theory of relativity and quantum mechanics, as it is the paradigm that currently underpins physics, astrophysics and cosmology (the study of the origin and eventual fate of the universe).

By the standard account of history of science a clear progression can be constructed. Progress is made through theories being articulated, tested and either proved or disproved. In the case of how the Sun shines there are a number of theories that are articulated and tested before the final, 'correct' answer is arrived at. When the final answer is achieved, the previous theories look a bit foolish, although they appeared to be highly plausible at the time.

Ancient wisdom actually provided quite a good explanation if one were to think in terms of the principles of formal scientific investigation: observations were made, a theory was formulated and evidence collected to support the theory. The nineteenth-century theories that replaced this were perhaps slightly more 'scientific', but both were disproved. In the case of the 'glowing ember' theory, the disproof came from looking at the problem from a different direction, and using the same evidence that had supported the theory in a different way. The disproof of the 'gravitational cooling' theory came not from inside the community of scientists investigating the Sun, but from geological evidence: in this case we have a clear example of the co-ordination between different parts of the project of science bringing about a palpable change in the status of a scientific theory. At the end of the nineteenth century, and after a number of centuries of dramatic scientific progress and discovery, the question of how the Sun works was still a puzzle: the mass of the Sun was too small to explain its workings by conventional chemical burning, and the age of the Earth meant that the Sun must have been 'burning' at a constant rate for a very long time. The solution provided by the application of Einstein's theory successfully answered the questions that physicists were asking about the nature of the Sun; it could explain the puzzle of the mass of the Sun and the age of the Sun, and thus achieved general approval from the scientific community, and subsequently from everyone else.

What we can see in this brief example through the history of science is the gradual adoption of a scientific framework of analysis for dealing with a problem in explaining the world around us: from an explanation that could not be tested, to one that relied on an amalgamation of experimental evidence with theory, to an explanation that combined experimental evidence with theory and with further evidence from other theories of the world, to an end point where a single theory – Einstein's theory – can explain all of the phenomena that surround us. We move from disparate explanations to a unified explanatory framework. But, interestingly, we have seen a decline in the amount of 'hard' evidence that is being brought to bear on the problem: ancient investigators were using evidence in the form of meteorites. Although they were wrong, they did at least have some tangible objects to support their theory. Similarly, the nineteenth-century 'burning ember' theory could use evidence from laboratory experiments investigating the rate of burning of various chemical substances in support. In contrast the gravitational cooling theory admitted no experimental evidence, and the nuclear fusion theory – the current orthodoxy – would appear to have been adopted solely from the compelling evidence supplied by a theory, Einstein's, that had received universal assent. This state of affairs is unsatisfactory for a model of scientific knowledge where the status of knowledge relies upon hard evidence being found to support a theory. Yet our astronomy textbook from 2010 clearly states that nuclear fusion is the 'true' theory. It will be necessary to return to this point. For the time being, we'll leave the question of how the Sun shines in favour of understanding just what this end point explanation means and signifies.

Standard history of science

Standard history of science presents us with triumphal progress. Scientific knowledge is cumulative and progressive: we used to know not very much, then a bit more and a bit more until now we pretty much know most of the big things about the natural world. It's worth noting that some commentators who adhere to the standard account, far from seeing this as being a good thing, will see it as being bad. Former senior writer at *Scientific American* John Horgan, for example, argues that science as a project of discovery is pretty much over, that we've found out everything there is to know and now we are just tidying up the last few details (Horgan 1996).

Being more specific, if we think about a scientific discipline such as physics we can say that ancient wisdom was often wrong about the world, but did get some things right. This was added to by the work of medieval and Renaissance scholars such as Copernicus and Galileo who extended our understanding of the nature of the physical world, and particularly the structure of the solar system. Then Isaac Newton came along and unified the laws of physics such that a great leap

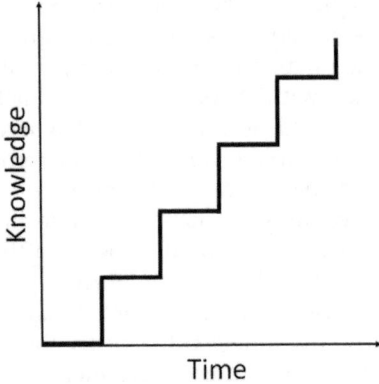

Figure 4.2 The progressive growth of knowledge

forward was made. Newton's theory produced a paradigm for physics with a set of theories that could be applied to all natural phenomena. In the nineteenth century, scientists such as Michael Faraday and James Clerk Maxwell extended our knowledge by building on previous work and applying it to new discoveries such as electricity and electromagnetic radiation. Then Einstein's work added to and extended the work of previous physicists. This model suggests that we are gradually, and stepwise, moving closer and closer to a full and complete explanation of the natural world, and specifically that Einstein's theories are, if not the final steps, very close to the final steps to achieving a complete understanding of the natural world. The standard account of scientific progress implies an end point where no further major discoveries can be made. Graphically we can display this on a Cartesian diagram, where we plot our knowledge of the world against time (figure 4.2).

It isn't just scientific knowledge that will fit this graph. A number of other ways of looking at the world, notably modernist social theories, imply exactly the same stepwise progressive model. One example would be the Marxist theory of historical progression through successive epochs. Substitute 'rationality' for 'knowledge' on the graph, and the fit is perfect, although we must note that Karl Marx and Friedrich Engels did not argue we were starting from a zero point, but rather held that we fell to that point through the invention of private property. Equally, the Whig interpretation of history, the standard liberal account of historical progress, broadly fits this model. This isn't just a coincidence: the reason these models match each other is because they all use a similar frame of reference for evaluating progress and for making sense of the world. Marx and Engels quite clearly describe and construct their historical materialism as a science which offers a rational analysis of the world. The project of science when describing itself is measuring only one thing – the amount of knowledge that we have – and is using a rational method to measure this. The Whig interpretation of history is using a rational method to measure progress, and focuses only on clearly progressive objects, such as human health, amount of technology available and rationalization of everyday life through, say, urbanization.

It is difficult to argue with such an account: many people in Western industrial societies are, after all, a lot healthier and have much more technology to help them in their everyday lives, the world around them appears to be much more rational, and science has clearly advanced fantastically in the last two centuries. However, a serious challenge to this idea of triumphal progress in the sciences, and by implication the triumphal progress of Western societies, has been made. The work of, and interpretations of the work of, Thomas Kuhn (1922–96) is most closely associated with a critical assessment of progress in the sciences: we will also note that a number of other writers arrived at similar ideas independently of Kuhn.

T.S. Kuhn and scientific revolutions

The previous chapter discussed the key elements of Kuhn's challenge to the standard account of scientific knowledge. Kuhn's central idea – of paradigms replacing each other through scientific revolutions, and that paradigms are incommensurable – is based on his revised history of science.

The idea of scientific revolutions taking place periodically through history is not new, and certainly not original to Kuhn's work. The main scientific revolution identified by historians, and discussed earlier in this chapter, is that of the period 1500–1700, which saw the foundation of modern science based primarily on the unification of physical laws in the work of Isaac Newton. The standard history of science is an account of rational progress, and scientific revolutions – the revolutionary overthrowing of one way of seeing the world in favour of another way of seeing the world – are seen as rational developments that are progressive and lead to cumulative growth of scientific knowledge (Popper 1981).

Kuhn's version of scientific revolutions is quite different from this account of linear progress. He considered scientific revolutions to be social revolutions as well; they relied upon social interactions to provide a critical mass of dissenting voices to bring about the necessary upheaval. In addition, a scientific revolution may not result in the progressive growth of knowledge. Prior to Kuhn, scientific revolutions were seen as dramatic upheavals in the ideas of a specific scientific discipline where a great leap forward was made that was an extension of existing knowledge. Thus in the standard account of scientific revolutions Einstein's theory provided a new paradigm for physics that was

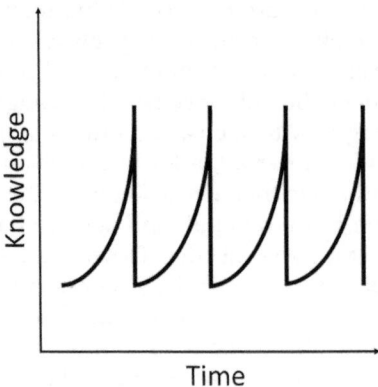

Figure 4.3 Discontinuous history of knowledge

a major step forward, but also a continuation of Newton's paradigm. In Kuhn's account a significant difference is noted: two paradigms can be *incommensurable*, and the language of the old theory may not be able to express the language of the new theory, and vice versa. This means that whilst any normal science will be cumulative, the science that takes place at the time of a scientific revolution – revolutionary science – is not necessarily cumulative. Quite the opposite: after a scientific revolution a huge chunk of the 'old' science will be abandoned as it is incompatible with the new science. A graphical display using a Cartesian diagram of scientific progress from the perspective of Kuhn's model (figure 4.3) is very different from that of the standard account.

What becomes visible is a discontinuous history. Progress is made inside a time frame which can be defined by the adoption of a paradigm, but then each subsequent scientific revolution knocks this progress back to a starting point similar to that of the previous paradigm. Kuhn's work is highly contentious, and argument still rages over whether his approach is correct. Central to the debate are two concepts: truth and incommensurability. The standard account argues that truth in scientific endeavour is something that is attained through the discovery of facts, and these facts can be built upon to generate new theories from which new facts can be discovered. However, as we have already seen, there is a problem with this realist conception of scientific knowledge: some 'facts' turn out to be not true and are replaced with new 'facts' that are considered to be true. The twist that Kuhn added, largely inspired by the work of Ludwik Fleck, was that different scientific communities at different times will identify different theories and facts as being true, and that these are incommensurable with one another. The question we need to consider is whether incommensurability between paradigms means that, whilst progress inside a paradigm can be discerned, there is no general progress towards a final goal in the project of science. Kuhn, in *Structure* is ambivalent about this:

> I do not doubt, for example, that Newton's mechanics improves on Aristotle's and that Einstein's improves on Newton's as instruments for puzzle-solving. But I can see in their succession no coherent direction of ontological development. On the contrary, in some important respects, though by no means all, Einstein's general theory of relativity is closer to Aristotle's than either of them is to Newton's. (Kuhn [1962] 1970: 206–7)

More recently, philosopher of science Ian Hacking has argued that the problems in interpreting Kuhn's work stem from a misunderstanding of what 'progress' means. Where our general, socially held idea of progress is 'moving towards a goal', Hacking reminds us that Kuhn argued that 'revolutions progress *away from* previous conceptions of the world that have run into cataclysmic difficulties. That is not progress towards a pre-established goal. It is progress away from what once worked well, but no longer handles its own new problems' (Kuhn and Hacking 2012: xxxiv). However, two points pertain. The first is that in the uptake of Kuhn's work it is clear that *Structure* is used as a tool to offer a critique of both scientific progress and social progress. The second is that many in the scientific community do not see scientific progress as a process of moving away from things, but rather a moving towards a goal. Here's Stephen Hawking: 'The eventual goal of science is to provide a single theory that describes the whole universe' (Hawking 1988: 10). Similarly, John Horgan in his book *The End of Science*: 'Science is not cyclic, however, but linear. . . . The biggest obstacle to the resurrection of science – and especially pure science, the quest for knowledge about who we are and where we came from – is science's past success' (1996: 24).

Box 4.2 Progress

Philosopher and historian of science Paul Feyerabend notes that the standard account of progress actually raises more problems than it solves. The story of progress and scientific progress tells us that 'we' 'know more' about the world than did our predecessors:

> [W]ho is the 'we' the critic is talking about? Is he talking about himself? Then the statement is quite obviously false – there is no doubt that Aristotle, on many subjects, knew more than he does. . . . Is the 'we' educated laymen? Again the statement is false. Is the 'we' all modern scientists? Then there are many things which Aristotle knew but which modern scientists don't know and from the nature of their business can't possibly know. The same is true when we replace

Aristotle by Indians, or by Pygmies, or by any 'primitive' tribe that has succeeded in surviving plagues, colonization and development. There are lots of things unknown to us Western intellectuals but known to other people. (Feyerabend 1988: 160)

Standard history of science tells us that the stock of facts and knowledge has increased, and Feyerabend has to agree that this point is probably correct: the sum total of the facts that now lie buried in scientific journals, textbooks, letters and hard discs does exceed the sum of knowledge from other traditions. 'But what counts is not number but usefulness and accessibility' (Feyerabend 1988: 160). Many research papers are never read, and even scientists in a specialist field will have no time to read all the relevant material. Feyerabend goes on: 'Most of the "knowledge" that is sitting around is as unknown as were quarks around the turn of the century' (Feyerabend 1988: 161).

Progress is a concept that we need to be careful in ascribing, even to a venture such as science which is avowedly about progressive accumulation of knowledge. Our society, and our science, rely on a story of progress to reassure themselves that they are headed in the right direction. Yet the work of Fleck, Kuhn and Foucault challenges not only the idea of progress in terms of the accumulation of knowledge, but also the idea of societal progress, i.e. the idea that our society is 'better' now than it has been in the past. Kuhn's work challenged the idea of progress in the formal sciences (see main text). In particular, the work of Foucault suggested that the idea of progress is simply not useful in making sense of our history. By showing that history is best understood as a series of discontinuous episodes, or *epistemes*, Foucault replaced the idea of progress with the idea of difference (Foucault 1970). Rather than being able to say that societies are better or worse than each other, Foucault suggests that all we can do is identify how societies are different from one another: we simply do not have the criteria available to us to judge in absolute ways. This point is similar to Fleck's observations about accepted facts inside thought communities; truth is almost always completely determined within a thought style (Fleck [1935] 1979: 100).The parallel to the work of Kuhn is clear: Kuhn's understanding of change through paradigm shifts corresponds to Foucault's idea of change through epistemes replacing each other (Radnitzky 1973: 380). This relativist position is now closely associated with a social constructionist understanding of the world.

Many scientists disagree with such relativist understandings of science, and the 'standard' account of science still represents the project of science as an inexorable rise towards greater and greater knowledge.

A history of sunshine 2

It is time to rejoin the story of how the Sun shines, this time from a post-Kuhnian perspective. We are now at the point where a 'correct' answer has been arrived at: by the 1920s the currently favoured theory was in place, and our examination of science textbooks confirms that there is little doubt as to the veracity of this theory in our present time. Notably, there are no other competing theories. The current theory – based on Einstein's work – presents scientists with a range of puzzles that need to be solved to prove and validate the theory. From the mid-twentieth and into the early twenty-first century we have seen a range of experiments designed to solve the theoretical puzzles presented by the theory of solar nuclear fusion. This fits closely with the Kuhnian model: the theory presents us with puzzles to solve, and normal science is a process of solving such puzzles. Scientists design experiments to solve the puzzles. The puzzles either confirm or disconfirm the theory. If they confirm it then we extend the theory to the bit of the natural world we have just looked at. If not, we think again, decide that the results we obtained are an anomaly and ignore them, or decide the results are an anomaly and reappraise the theory.

In the case of the 'how does the Sun shine' puzzle we have a pretty clear-cut case of what needs to be done to solve it. Einstein's theory predicts that the Sun works by a process of nuclear fusion, hydrogen atoms fusing together to produce helium, and a lot of energy is released in the process. We want to test this out – solve the puzzle of how the Sun works – so we need to design experiments that do this.

We could just measure the energy that comes from the Sun. Doing this will confirm that the Sun radiates a certain amount of energy, and we can use our theoretical prediction of how much energy should be generated by that mass of gas to test the theory. If they match up then we might want to say we have solved the puzzle. But we wouldn't have, or only partly. It could just be coincidence that the two figures match up. Think back to the 'burning ember' theory – that may have matched the two figures (mass and energy released), but we know from other scientific interventions into other aspects of the phenomenon of sunshine that the theory was not correct. Much better than just observing would be a process of testing the theory by designing an experiment.

To solve the puzzle, we need to look more closely at the specific aspects of the theory and find other points where we can match experimental results against theoretical predictions – and it is here that we get our experimental breakthrough. The process of fusion is, as noted, quite complex, and involves three distinct steps, according to the current theory. Summarizing the whole process, four protons (hydrogen nuclei) fuse together to make a helium atom, and two gamma rays and two neutrinos are released (see figure 4.4 on p. 125). If we could count these reaction products (i.e. the gamma rays and the neutrinos) we could verify the theory. We can 'see' the gamma rays coming from the Sun, but they may not be the product of the reaction we are trying to test for – gamma rays come from lots of places and can be made in lots of different ways (e.g. by radioactive decay on Earth or from distant supernovas thousands of light years away). It will not do us much good to count gamma rays in a world that is saturated with them from many different sources – we probably would not be able to 'sort' the solar ones from the other kinds. That leaves the neutrinos: if we could count the neutrinos coming from the Sun we could verify the theory and this would tell us if the reaction is actually taking place. So we need an experiment to measure the solar neutrinos, and with the results we can compare these to the predictions that our theory is making. If they match up – bingo, we've proved that the Sun works by nuclear fusion as predicted. If they don't match up, we need to reappraise our theory, or do the experiment again, perhaps in a more sophisticated way.

Neutrinos, being neutral particles that very rarely react with other particles, are extremely difficult to detect. Nevertheless, in 1967 a detector was built – 400,000 litres of dry-cleaning fluid in an abandoned gold mine in South Dakota. The experiment, designed by Ray Davis of Brookhaven National Laboratory, was called 'Homestake' and it ran for about twenty years. The experiment needed a very long run time to collect a sufficient amount of data for robust conclusions to be drawn. Subsequently a number of other solar neutrino detectors have been built to try and confirm or deny the Homestake results.

Given the predicted number of neutrinos coming from the Sun (many trillions per second), the Homestake experimenters calculated that their detector should locate an event about every day – which shows you just how unreactive neutrinos are, and why the experiment took so long. However, in contrast to the prediction, neutrinos were captured only about once every three days on average. There were only one-third the number of neutrinos coming from the Sun that the theory predicted. This discrepancy between the actual number and predicted number of observations is called the solar neutrino problem. (Collins and Pinch (1998: ch. 7) provide a comprehensive overview of the solar neutrino problem from the start of the Homestake experiment up to 1998.)

This bears some further investigation. The theory has made a prediction, it is tested and it is found that the results don't match up – there is an anomaly. What should be done? Perhaps the detector isn't working? There have been a number of more recent attempts to build better detectors to try and find the lost neutrinos, and the Homestake results have been confirmed by these later detectors.

Perhaps it is the big theory that is wrong? That leads to dangerous territory – after all, the whole paradigm of physics is based upon Einstein's equations, and knocking them down would have huge con-

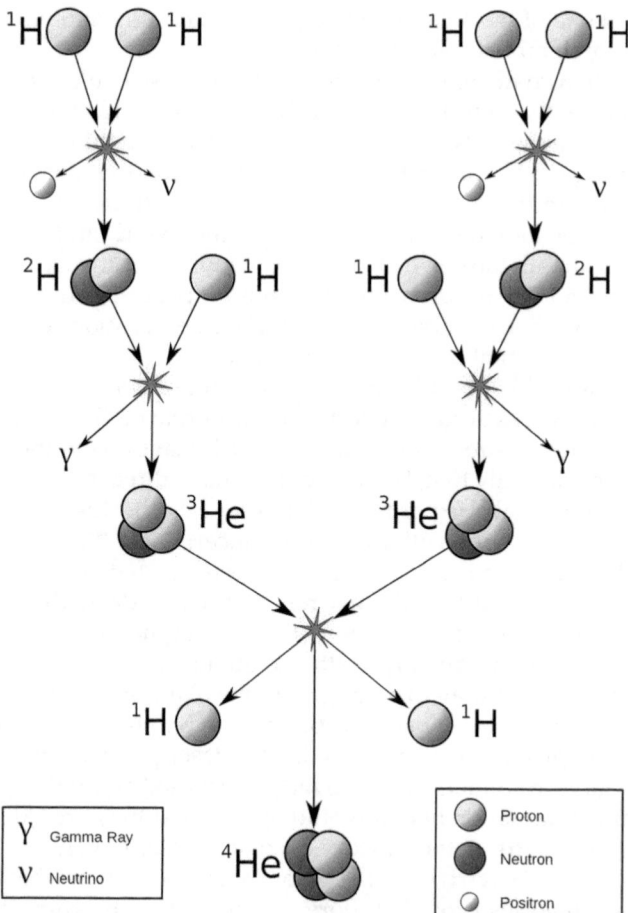

Figure 4.4 Solar fusion

sequences. Einstein's theory does a good job of explaining many other natural phenomena in the universe. Yet the possibility that Einstein's theory is wrong must be considered. This is not a trivial point: the solar neutrino problem goes to the heart of the paradigm of physics. It isn't just the case that if the problem can't be solved we would have to reappraise our understanding of how the Sun and, of course, all other stars in the universe work. It would also suggest a fundamental reappraisal of the main elements of particle physics was necessary; if fusion doesn't work like this, then what about all the other predicted interactions between non-observable particles? And what about all of Einstein's other predictions? However, this could be seen as the least likely outcome of the solar neutrino problem: finding another solution would be preferable for most observers.

Rather than challenging the whole paradigm, we could look at some of the smaller theoretical aspects of the process, such as the nature of neutrinos – maybe we've been looking for the wrong things, or looking in the wrong way? The problem here is that if we do manage to make any progress in this area, it will probably be at a theoretical level: our *experiments* for looking at neutrinos, how they work and where they come from, are already in difficulties. Changing our theory may make more sense in that we can make neutrinos on paper fit our predictions better, but that doesn't alter the fact that our observations of neutrinos are giving results that need to be explained.

Note that at this point in the story we actually do not really know how the Sun works according to formal scientific methodology. There is a theory, but it is resistant to proof, which leaves us in the realm of speculation. Not only that, but we actually have direct factual evidence from the Homestake experiment that suggests the theory is wrong. How do scientists cope with this sort of uncertainty? 'For the moment, many physicists and astronomers are *betting* that we understand the Sun just fine and that the discrepancy has to do with the neutrinos themselves' (Bennett et al. 1999: 476; my emphasis). This quotation is from the same astronomy textbook quoted earlier. We've moved from triumphal certainty to betting. And the blame here is being placed firmly on the neutrinos, not on the big theory. No challenge to the main paradigm is proposed or discussed. The description in the textbook implies that astronomers are not overly concerned about this problem, and from the quotation here it would appear that they are exhibiting a high degree of **faith** in their theory of solar nuclear fusion.

So, we have now arrived at the paradoxical position in which the old discredited theories provided more certainty than the current theory. With the old theories it was 'obvious' that the Sun was a big ball of fire burning away in the sky, and there were even bits of hard material evidence (meteors and meteorites) at which people could point. We didn't have a situation where people were saying 'who knows?'. With our twentieth century situation, we simply couldn't prove how the Sun shines – we had less certainty. This provides a good example of Kuhn's theory of incommensurability, which we can plot on a graph (figure 4.5) similar to that in figure 4.3 on p. 120.

(1) Ancient thought was confident that the Sun was a big ball of fire that moved around the fixed Earth: confirmation came from collecting evidence in the form of meteors that fell to the ground. But when it was realized that the Sun was at the centre of the solar system, not the Earth, this theory was plunged into crisis and a new theory, eventually, emerged.

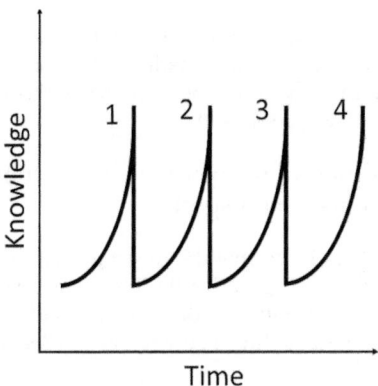

Figure 4.5 Discontinuous theories of how the Sun shines

(2) Early modern theories of how the Sun shines were based on analysis of how various substances combusted and remained incandescent. Evidence from other sources indicating the age of the Earth to be about 6,000 years coincided with this. However, dramatic results from fossil records put paid to these theories.

(3) Late nineteenth-century theories focused on the new knowledge that the Sun was made of hydrogen and helium and was very large. The theory of gravitational contraction answered all the unsolved questions left by previous theories, but ultimately failed because it could not explain how the Sun could shine for so long.

(4) Einstein's theory makes predictions about energy released by nuclear fusion, and this becomes the next theory to be adopted. Again, the theory can answer all the questions left unanswered by previous theories, but is still lacking definitive proof.

However, as I suggested earlier, it is almost inconceivable that most people in contemporary society would think that there is less certainty about how the Sun works now than in, say, 1850. Our technoscientific society does not admit the possibility of there being major holes in our knowledge of large parts of the operation of the natural world.

The solar neutrino problem has been 'solved', but only recently (2001), and possibly in a way that many will not find particularly satisfying. As one would expect, the 'solution' confirms the paradigm-inspired theory of nuclear fusion:

> At last physicists have solved a problem that's been plaguing them for three decades. Why does the Sun seem to emit fewer neutrinos than it should? *It's simple.* Neutrinos can change from one form to another. Until now physicists had begun to think that either our understanding of the structure of the Sun was wrong, or that neutrino detectors did not work

> properly. But this week, a team of researchers from Canada, Britain and
> the US announced that neutrinos made in the Sun can change to other
> types en route to Earth. The other kinds are harder to spot, and so fool the
> detectors (Cho 2001; my emphasis)

The more recent edition of the astrophysics textbook I've used in this
chapter is slightly more circumspect:

> Scientists are now convinced that the missing solar neutrinos were going
> undetected. . . . Strong evidence for the idea that solar neutrinos change
> type comes from a detector in Canada called the Sudbury Neutrino Obser-
> vatory, which can detect all three neutrino types. (Bennett et al. 2010: 481)

A happy ending, and the story has an even happier ending for the scientists involved in the long and painstaking search for the elusive solar
neutrinos. In 2002 the Nobel Prize for physics was awarded to Ray Davis
and Masatoshi Koshiba for their work on the solar neutrino problem.
Full details of the experiments and a transcript of Davis's acceptance
speech are available on the Nobel Prize website at: http://www.nobel.
se/physics/laureates/2002/index.html. Solar neutrinos are an interesting topic for popular science too; Frank Close's book *Neutrino* makes
Ray Davis its hero, describing Davis as 'the first person to look into the
heart of a star' (2010: vii), and the 'thrilling detective story' of the hunt
for solar neutrinos is recounted in Ray Jayawardhana's 2013 book *The
Neutrino Hunters*. But is this really the end of the story?

Exoteric and esoteric histories of science

How and why are histories of science written? This is not an easy question to answer. Apart from anything else there are a lot of histories of
science available to a reader: searching online with the string 'history of
science books' returned 390,396 titles from amazon.co.uk on 3 July 2014;
the number one hit was John Gribbin's *Science: A History 1534–2001*
(2003), discussed earlier in this chapter. Such a large number precludes
a detailed analysis of this genre. Rather, what I hope to do is illustrate
four ideal typical positions in a field that we can characterize as being
split into two main divisions, defined by the thought communities that
readers and authors are members of. History of science writers fall into
two broad categories: formal scientists who write history – often of
their own specific sub-discipline or research specialism – and historians
who take formal science as their specialist topic (as opposed to, say,
economic history or political history). We can see from even a cursory
overview of their output that they are writing for either a specialist,
esoteric audience or a non-specialist, exoteric one. As there are grey
areas between these extremes, I propose that we can view these texts on

a continuum from esoteric to exoteric, and that history of science texts as a whole can be arrayed on two continuums: one where the authors are formal scientists, the other where the authors are trained historians. As with all ideal types, there will be significant deviation from these positions – and these deviations serve to strengthen the solidity of the ideal types.

We have two continuums: historians' histories of science and formal scientists' histories of science, and four polar positions:

- histories of science for the esoteric history of science thought community,
- histories of science for esoteric formal science thought communities,
- exoteric histories of science for the general public written by historians,
- exoteric histories of science for the general public written by formal scientists.

We can thus identify at least, but principally, four audiences:

(1) professional scientists who are interested in the history of their discipline or sub-discipline;
(2) professional and trainee historians of science who are interested in how the history of science can be written, and what its key features are;
(3) general, 'lay' readers who want authoritative, insider (i.e. scientist-written) histories of science to inform them about where the roots of contemporary science lie;
(4) general, 'lay' readers who want a contextualized history of science, or aspects thereof, to help inform their understanding of the social, political, economic conditions surrounding the emergence of formal science.

It is likely that if we were to carry out empirical audience research we would find that audiences (3) and (4) are coterminous, i.e. the audience is the same regardless of the authorship of the text.

However, and more significantly, of the four ideal typical positions, at least three (all except historians' esoteric histories of science) express a very similar, and strongly normative, account of what science itself actually is. It is interesting to note that whilst the esoteric histories of science written by both scientists and historians of science have little or no appeal to a general audience – indeed, positively eschew such an audience – both types of exoteric science are vying for the same audience. And, perhaps surprisingly, we see some remarkable overlaps here. It isn't the case that these exoteric histories are dissimilar from one another: quite the opposite. These general histories of science, or aspects of formal science, present a similar, and predictable, picture of science. Most history of science texts, and particularly those written by scientists or exoteric histories of science, present science as if it had a history independent of the rest of the world, including society. There are a huge number of books such as this – Hawking (2002), Asimov (1989), Gribbin (2003), Bryson ([2003] 2004) and so on. Science has a separate history that largely ignores the external, non-scientific world, and admits of no external influences. These histories are purely internal accounts, except where events absolutely force themselves on the narrative, as in the rapid development of physics following the inception of the Manhattan Project: science is not placed in any wider context, particularly not a social one. Only the esoteric histories of science written by historians, and then only some, challenge or problematize the representation of science as external formal knowledge that is progressive, created by individuals largely independent of any external factors or even ideas, and something that is not connected in any way to the 'lay' public (Erickson 2010).

From a Kuhnian perspective history needs to be written *inside* a paradigm, i.e. needs to be discipline specific, and the esoteric histories certainly are so. In contrast, general and popular histories of science (such as Asimov's or Gribbin's discussed earlier) look at the whole of formal science across a large sweep of time and usually constructing a triumphal march-of-progress account. But these are also internal accounts, considering science to have **only** an internal history and almost always excluding reference to external events or factors that may impinge on the development of science. And both of these forms of historiography, the exoteric and the esoteric, share a feature: a prominent lack of detail. We saw in chapter 2 that scientific research work is very detailed, time-consuming, painstaking and difficult. Inside one lab there will be a general increase in knowledge outputs over time, but not either a linear one or necessarily a triumphal one. The details of the daily grind of formal science work, the failure of experiments and the exigencies of laboratory life are largely written out of these large-scale historical accounts. In esoteric texts where the minutiae of the work of formal scientists is considered we once again find a lack of any kind of contextual or explanatory perspective. For example, Peter Galison's *Einstein's Clocks, Poincaré's Maps* (Galison 2003) is meticulous, beautifully written, fascinating, and written from an insider's perspective and understanding of the physics thought community. And yet this history of the synchronization of time in the nineteenth and early twentieth centuries, with a focus on railways and the telegraph, which claims to be 'an inextricable mix of social history, cultural history, and intellectual history; technics, philosophy, physics' (ibid.: 328), contains no mention *whatsoever* of industrialization, social change or capitalism.

Instead we have a history of the personalities, experiments, machines and academic institutions that brought about time synchronization in the late nineteenth century, and this shows how conceptualizations of time and synchronicity change between different locations and groups of practitioners.

Most histories of science serve a dual purpose. Although using quite different content (very specialized formal knowledge versus general, 'compiled' scientific narratives), both represent science as isolated through, on the one hand, a reinforcement of the 'specialness' and superiority of science due to its method, attitude and project and, on the other hand, emphasis on the role of the lone genius, the driven scientist who leaves society to create purer scientific knowledge. Both serve to present an account of science that can be misleading. As Fleck notes:

> The history of science also records cases of independent – one might say personal – exploits. But their independence is only characterized by an absence of collaborators and helpers, or possibly of pioneers; that is, it manifests itself in the personal and independent concentration of histori-cal and contemporary collective influence. In a manner corresponding closely to personal exploits in other areas of society, such scientific exploits can prevail only if they have a seminal effect by being performed at a time when the social conditions are right. (Fleck [1935] 1979: 45)

It is difficult to imagine any other kind of history that would or could be written in this way. Political histories that considered only the changing fates and faces in political parties with no reference to, say, economic trends and tendencies in a given society, or military histories that looked only at the actions of armies on battlefields and did not address or even discuss the causes and consequences of wars, would not, I think, be read by many people. We wouldn't even call such accounts 'histories'; they would be called 'chronologies'.

There are a number of accounts – generally more recent accounts – that have attempted to challenge these 'standard' histories. Rather than attempting to explain how a specific discipline emerged, or looking at the growth of knowledge of a specific phenomenon, some histories examine in great detail specific scientific practices and trends from perspectives inspired by philosophy and sociology of science. In *Image and Logic* (Galison 1997) Peter Galison presents a detailed account of the development of particle physics through the twentieth century, focusing not on the theories, objects or personalities that are associated with this discipline, but rather on the material cultures that scientists were involved in constructing to investigate subatomic particles.

Although this approach – focusing on the equipment rather than the theories – may appear to be marginal, the results are useful for social investigators of science. For example, the emergence of 'big science' in the USA as a consequence of the World War II Allies' Manhattan Project to design and build the atomic bomb is already well known, but the adoption of modern management theory, modes of organization and attitudes is an interesting point to identify. Such conjunctions of thought – between capitalist enterprise and the physics laboratory – could not be identified from an examination of the development of theory, or by focusing on the leading scientists in a specific area. Indeed, 'standard' history of science would obscure such points.

Andrew Pickering's *Constructing Quarks*, a study of the history and sociology of high energy physics, also uses a more sociological frame of reference to make sense of a historical narrative (Pickering 1984). Pickering notes that, in general, histories of science represent 'scientist's accounts' and, notably, are written in such a way that the scientists involved in the production of the knowledge in question do not appear as genuine agents:

> In the scientist's account . . . [s]cientists are represented rather as passive observers of nature: the facts of natural reality are revealed through experiment; the experimenter's duty is simply to report what he sees; the theorist accepts such reports and supplies apparently unproblematic explanations of them. One gets little feeling that scientists actually *do* anything in their day-to-day practice. (Pickering 1984: 7–8)

Sociologically inflected histories, such as Pickering and Galison's, are attempts to introduce the scientist as agent into the history being written. Without doing this the history of science continues to replicate the dominant and standard account, where science is a triumphal progress, and is also a consolidation of the dominant scientistic attitude towards knowledge.

However, despite these efforts at producing 'alternative' histories of science, the micro-examination of the internal workings of a scientific sub-discipline, for all that it uses a sociologically inflected frame of reference, is still largely an esoteric account that does not challenge the fundamental assumptions of the project of science. Two issues – that of gender and that of colonialism – are embedded in standard histories of science, but simply reproducing the standard esoteric approach cannot begin to challenge these underlying assumptions.

Historian of science Patricia Fara, as well as producing a comprehensive alternative history of science (2009) that squarely challenges the standard account, is a leading voice in uncovering and highlighting the role of women in the history of science. She points out that women have been active in scientific endeavours throughout history, although men deemed their role in science to be 'unnatural'. Despite the significant achievements made by women in science they have been systematically excluded from written histories of science (Fara 2004). Fara's histories of science bring to light the contributions women made, but also show the systematic exclusion and discrimination that women have faced in the past and still face today. The final section of her history of science for younger readers ends with a short overview of the continuing discrimination that women currently face in science environments (Fara 2005).

Zia Sardar notes that despite much of Western science being based on fundamental discoveries made in non-Western societies the history of science is written from a wholly Western perspective.

> The conventional (Western) history of science . . . does not recognise different types of civilization or cultural sciences. It has represented Western science as the apex of science, and maintained its monopoly in four basic ways. First, it denied the achievements of non-Western cultures and civilizations as real science, dismissing them as superstition, myth and folklore. Second, the histories of non-Western sciences were largely written out of the general history of science. Third, it rewrote the history of the origins of European civilisation to make it self-generating. . . . Fourth, through conquest and colonisation, Europe appropriated the sciences of other civilisations, suppressed the knowledge of their origins, and recycled them as Western. (Sardar 2000: 53–4)

The post-colonial challenge to the history of science has grown in strength in recent years. Post-colonial STS seeks to 'reclaim the history of non-Western science and expose the Eurocentrism of Western science' (Sardar 2000: 55). Such studies have revealed strong links between the development of Western science and the development of Western colonialism: science was used as a tool by the colonial powers to enforce their rule. It's a point that brings us back to the main argument of this chapter: that unless we can see science in the context of much wider social, economic and cultural happenings we will simply not understand science and its place in our society.

Conclusion

The history of science, like formal science itself, presents us with many problems in terms of definition and understanding. We have seen how the idea of science has shifted and changed from Hobbes's time, when it was something abroad and demotic, to being something that is isolated and confined, a thing with an essence to itself. At the same time, histories that have been written of science preserve this conception of science as isolated and separated; even Kuhnian critiques of the standard history of science have maintained this isolationist approach to understanding science across time, preserving the idea that it has an essence that is transcendental. Bizarrely, almost all histories of science hide the detail of what occurs during, and the actual personnel involved in, making scientific knowledge and the wider socio-economic and cultural context that formal scientific knowledge emerges from and into. Yet despite these deficiencies the standard history of science has very wide acceptance in our society, and communities of academic historians of science cleave to this way of 'doing history' as they construct their esoteric accounts. Why is this the case?

History of science, patently, will not help us to make sense of this. We need to move into a much more sociological and critical space to begin to unpack this situation. Sociology of knowledge is particularly helpful here:

> The principle thesis of the sociology of knowledge is that there are modes of thought which cannot be adequately understood as long as their social origins are obscured. It is indeed true that only the individual is capable of thinking. . . . Nevertheless it would be false to deduce from this that all the ideas and sentiments which motivate an individual have their origin in him alone, and can be adequately explained solely on the basis of his life experience. (Mannheim [1936] 1960: 2)

Mannheim's starting point is not only that knowledge is formed in social contexts and reflects social conditions, but also that the knowledge we have of the world will affect how we make sense of the world and how we will act in it. Michel Foucault's histories take a similar approach, and to great effect. In particular, Foucault shows that knowledge is always attached to power and vice versa, and that our sense of self and identity is constructed from discourses that bring together knowledge and power. In examining the history of science we need to do something similar and consider what the social conditions are that are leading to this form of knowledge being produced, and what the consequences are for a society that deploys this form of knowledge. This is clearly a large undertaking, and beyond the scope of this book (although well worth pursuing, I think). However, we can make a start and there are clues we can follow up. Most Western histories of science ignore two things: colonialism (discussed earlier) and gender – Galison's *Image and Logic* is one of the few historical accounts that considers the gendered division of labour in high energy physics. In almost any other area of academic historical investigation these things would be at least on the agenda. Why are these absent in the history of science?

Scientism emerges in tandem with the development of capitalism, and is a form of knowledge that helps to sustain the structures of power we find in capitalist societies. The scientific universalism that it propagates makes claims to have access to the truth and to produce the only form of legitimate knowledge. To support this, scientism deploys a history of science that makes similar claims: science is progressive, objective and taking us towards a goal. It also reinforces the exclusion of morals and values from legitimate knowledge: truth is truth and scientific universalism does not concern itself with what is right or wrong, good or bad. But as we will see in chapter 5 formal science communities exhibit a gendered division of labour and a hierarchy that systematically favours men over women. Similarly, scientists from ethnic minority backgrounds and science from non-Western and industrialized countries are likely to receive less reward and acclaim. However, scientific universalism implies that those in positions of power and influence are morally entitled to be there as they have risen through an (alleged) meritocracy. The social structural conditions that prevent people from achieving high position in this meritocracy, such as sexism and racism, 'were basically eliminated from analysis' (Wallerstein 2006: 78). We are in a situation where the legitimated knowledge in our society excludes, or rather demotes, questions of morality and fairness. Asking such questions of scientism and scientific universalism, and by extension formal science itself, can lead to charges of anti-intellectualism. Yet these questions are the most important for any society facing difficult choices in an increasingly globalized world.

The history of science is an essential component of the scientism that predominates in our society. It tells a story of triumphal progress, of great technological advances, of the emergence of a powerful, neutral and objective scientific method, and of the benign meritocracy that allows only the best individual scientists to reach the top. It reinforces the idea of individuals being responsible for themselves, setting their own rational and material (but not moral or ethical) goals and being in competition with other individuals; a neoliberal vision of contemporary life (but also a Hobbesian one). We will turn to the relationship of individuals to their scientific communities, and that of scientific communities to the rest of society, in the next chapter.

Further reading

There are a lot of general histories of science, and this chapter used a few as examples of the 'standard' history of science. Histories that don't simply reproduce the standard account but place science in its social, economic and political context are much less common, but much more interesting and useful. Here are two:

Agar, J. (2012) *Science in the twentieth century and beyond*, Cambridge: Polity.

Fara, P. (2009) *Science: a four thousand year history*, Oxford: Oxford University Press.

John Agar's book grounds the formal science being carried out in the context of the 'working worlds' that society is oriented towards. This brings the exoteric and esoteric into direct contact and provides grounded explanations for why developments occurred in the way that they did. Patricia Fara's comprehensive history of science is an excellent example of how science can and should be placed in context to understand it, and to challenge the myths surrounding history of science.

5

Scientists and Scientific Communities

Community can be the warmly persuasive word to describe an existing set of relationships, or the warmly persuasive word to describe an alternative set of relationships. What is most important, perhaps, is that unlike all other terms of social organization (state, nation, society, etc.) it seems never to be used unfavourably.

Raymond Williams, *Keywords* (1983)

We encounter the words 'scientific' and 'community' in conjunction with each other in a number of different places, and with a range of connotations, but all of these, as Williams notes in the epigraph to this chapter, are favourable. The exoteric language-game of science that we can see in many media representations creates an object – the scientific community – that is the product of a number of discourses that articulate scientism in our society. The idea that scientists are all broadly similar and can be lumped together in a single category, and that this category can be seen as a single and real institution, *the* scientific community, is strongly held in our society. There is a strong set of ideas about who scientists are, what they do and where they are. In this chapter we will begin by looking at how 'the scientific community' emerges as a discursive construction from the largest thought communities in our society and consider the implications this has for our understanding of science. This collectivization of scientists into a single community is a product of imagination, but this is not a simple process as described by Benedict Anderson (1991), where it is the members themselves who are doing the imagining. In the case of the collective scientific community we can see two processes at work. The first is the public construction of the scientific community; the exoteric imagining of what 'all scientists' are like and do. The second is the way that formal scientists themselves imagine their community. This occurs in two ways. Firstly, prominent formal scientists and their representatives make claims for all formal scientists to be seen as a singular grouping – again, *the* scientific community. We can see this large community in contradistinction to the esoteric thought communities of formal scientists, communities that are made through the practices and communicative interactions of their members as well as being constructed discursively and imagined.

The idea of 'real' versus 'imagined' communities (Anderson 1991), 'grounded' versus 'projected' communities (Delanty 2010), is not particularly helpful to us here, or in other studies of communities (Delanty 2003: 170), although the terminology is seductive. It would be nice to say that some communities are real by virtue of their location in a place or space and others are imagined and thus have no actual existence other than in our minds. However, two things prevent us from adopting this classificatory scheme here. The first is that sociological accounts of community often focus on connections to place, but many of the forms of collective interaction we have in contemporary society do not have a physical location, taking place in virtual environments and across large distances. Not only that: these forms of 'community' may engender the traditional idea of 'belonging', but such a sense of self can often be resisted or rejected in favour of quite different forms of belonging. Secondly, if we reject the idea that community is about place – and we must do that if we are considering scientific communities; if they do have one 'real' characteristic it is that they have no connection to place – then we must see community as a construction made by either self-identifying members or those who are observing something from the outside, then describing and thus constructing a community. Both of these points come together to make it vital that we look at how communities are constructed discursively. When we consider 'the scientific community' in this way we can see that it exhibits features that are an articulation of the scientism and standard account of science that are dominant in our society. And when we begin to look at the structure of the scientific community, in terms of its composition and hierarchy, we will see that it reflects the androcentric and patriarchal trends in our society; no surprise, perhaps, were it not for the obvious conflict this presents to the avowed and oft-asserted objectivity and neutrality of science.

Representing the scientific community

The general public, that is, those people who are not particularly familiar with the theory, methods and procedures of formal science and scientific knowledge, often refer to the practitioners of science as 'scientists'. This is an interesting term, one that immediately tells us something about the group being referred to and the individuals that it contains. It implies that we all know who scientists are, where we can find them, what they do, even what they look like, what their attitudes towards the world are and what sort of cultural preferences they have. But it is notable that those who *are* familiar with the theory, methods and procedures of the formation of formal scientific knowledge also refer to themselves as 'scientists'. We have a persistent societal image of the practitioners of scientific work – scientists – as a collective, and escaping from that image will be difficult. The image itself is unhelpful for understanding what science, scientific work and scientific knowledge are, and it is worthwhile looking at how and why it is that scientists are collectivized in the public imagination and cultural representations.

A quick survey of two UK newspapers (the tabloid *Mirror* and broadsheet *Guardian*) reveals just how prevalent in the press stories about scientists are. Both newspapers had five stories that included the word 'scientists' on Friday 4 July 2014, the day before I wrote this, and both had many thousands of hits in their archive, suggesting that the day's tally is not atypical. A few, selected, examples of headlines and their immediate straplines will give a flavour of the sorts of stories we can find:

[1] Breast cancer risk could be reduced by statin drug taken to help prevent strokes.
  Scientists found a link between cholesterol and the disease, which affects some 50,000 women a year. Breast cancer risks could be cut by taking statin drugs which lower cholesterol, say scientists. (Andrew Gregory, *Mirror*, online edition)

[2] Earth-like planet boosts chances of finding aliens . . . and it only takes 3,000-light [*sic*] years to get there.
  It seems we may not be alone . . . astronomers say. . . . The chances of finding alien life in space have been boosted by the discovery of an Earth-like planet orbiting just one of two stars, scientists claim. (Sean Garnett, *Mirror*, online edition)

[3] Facebook denies emotion contagion study had government and military ties.
  Researchers say the study was not funded by Minerva Research Initiative, which engaged scientists in national security issues. (Samuel Gibbs, *Guardian*, online edition)

[4] 3D printed organs come a step closer.
  Australian and US scientists make major medical breakthrough in printing vascular network. (Melissa Davey, *Guardian*, online edition)

[5] Another scare story? Don't panic, just follow the money.
  Politicians and scientists have a vested interest in propagating fear: it's the one superbug there's no known antidote for. (Simon Jenkins, *Guardian*, print edition)

These five examples are chosen from the many available that illustrate how the UK media deploy the term 'scientists'. There are, of course, other usages available, but the aim here is to give a flavour of the commonality of this usage, and some of the key features of the deployment of this term. Firstly we should note that what we are really seeing is two separate forms of usage. The first, usage 1, is a process of hiding, the removal of the specific disciplinary category that an individual would describe themselves as being, for example botanist, biochemist or palaeontologist, and its replacement with a generic term 'scientist'. This first usage of 'scientists' therefore designates a group of individuals involved in a common research effort or discipline-specific project, as in example (1): the *Mirror* article is, presumably, referring to research carried out by cardiologists and biochemists. For such articles in the press, the identifier of 'scientists' would appear to be sufficient to provide the reader with an idea of who is being talked about, and no further identification is usually provided (although example (2) provides the reader with both 'astronomers' and 'scientists'). The accuracy or otherwise of this idea that the reader constructs is not the point here, for the moment at least. However, it is unlikely that we would do this with other occupational categories: describing both Miley Cyrus and Dame Kiri Te Kanawa as 'musicians' is accurate but misleading in the same way that describing palaeontologists and astrophysicists as 'scientists' is accurate but misleading.

Usage 2 is slightly different. Where usage 1 hides the specificity of the individual's work and research and replaces it with a generic term, usage 2 elides the specificity of the individual's social situation and replaces it with a construction of community and collective ethos. Example (5) suggest that all scientists are part of a large grouping, a community that can have an agenda over and above what their daily work requires of them. It also implies location in that this is a community that has a unified voice that emanates from a single locus where scientists can be consulted by the media. As far as the media is concerned, scientists – by usage 1 – are pretty much interchangeable with each other, regardless of discipline or location, and, by usage 2, are all in one big grouping. Both of these forms of usage illustrate how society, via the media, is constructing an image of scientists and a community of scientists, *the* scientific community.

These two forms of usage suggest that there are two different ways of imagining scientists in wider groupings that are visible in society: as aggregated according to disciplinary boundaries, and as aggregated as a single mass of scientists. It is notable that these two forms of definition, the general and the particular scientific communities, correspond broadly to the two main academic conceptions that have structured sociological studies of the scientific community.

Sociology of the scientific community

The first major conceptualization of scientific community comes from the work of Warren O. Hagstrom, who in his seminal 1965 work *The Scientific Community* presents an account of the relationship between individual scientists and the wider community they form together. It is this wider community that regulates their work, assigns status to individuals and groups, and articulates the commonly held values of scientists. Hagstrom's book is the first widely distributed account of the formation and structure of scientific communities, although he did not invent the term 'scientific community'. The antecedents go back some years: Baldamus argues that (1977 fn. 1) it was Ludwik Fleck in *Genesis and Development of a Scientific Fact* (Fleck [1935] 1979) who first coined the phrase. Hagstrom's thesis describing the organizing principles and mechanisms of the scientific community is 'that social control in science is exercised in an exchange system, a system wherein gifts of information are exchanged for recognition from scientific colleagues. Because scientists desire recognition they conform to the goals and norms of the scientific community. Such control reinforces and complements the socialization process in science' (Hagstrom 1965: 52). Conformity is rewarded and commitment to higher goals is also reinforced. These higher goals include a commitment to the scientific community itself; this extends beyond any specific collection of peers, and beyond any disciplinary boundary. In considering conformity and social control Hagstrom does note that scientists are also employees, but here largely rejects any possibility that commitment to production inside science is a result of individuals' feeling bound by contractual obligations (ibid.: 54). For Hagstrom, the productive nature of science stems from the open and egalitarian nature of the scientific community and its organization around gift-giving principles.

Hagstrom's thesis emerges from primary data collection he conducted, largely based on in-depth interviews with academic scientists in a range of disciplines working in US universities. His empirical work is an extension of the Mertonian theory of the institution of science (see box 5.1), an institution that is governed by imperatives that express the values of the members, and can be seen as an empirical 'testing' of Merton's theory.

Indeed, Hagstrom himself makes generous reference to the work of Merton as being of great influence on him, citing him frequently in the main text, and making special mention of the debt he owes him in the acknowledgements. Hagstrom, perhaps surprisingly, offers no clear definition of 'the scientific community', but as he makes no reference to the possibility of a range of different communities co-existing we can assume that his scientific community is unitary, and has a hierarchical structure of disciplines and individual members: both are

---

Box 5.1 The institutional imperatives of science

Robert K. Merton (1910–2003) identified four imperatives that characterize the ethos of science (Merton [1942] 1957: 550–61).

(1) Universalism: truth claims are to be subjected to pre-established impersonal criteria. All claims are approached using similar methods (universal methods) – and all claims are as likely as others until proven otherwise. For Merton, this expands to give the ethos of science a view of careers: these should be open to talents, not prejudice. Universalism rejects versions of science that are particular, such as the science of Nazi Germany which was based on 'Aryan science'.
(2) Communism (here Merton means the common ownership of goods): the substantive findings of science are a product of social collaboration and are assigned to the community. Property rights in science are whittled down to a bare minimum by the ethics of the scientific community. Secrecy is the antithesis of this norm; communication of findings is a must. The only things scientists own are the esteem and recognition due to them by the wider scientific community.
(3) Disinterestedness: there is competition in the field of science, but this is not at a level of discrediting others' findings without good scientific reason. Disinterestedness also requires scientists to scrutinize their work and that of peers in such a way that the main winner is 'the truth', not any individual's personal beliefs.
(4) Organized scepticism: the suspension of judgement until the facts are at hand. The scientist suspends their common sense – thus allowing the hidden truth to appear.

---

ranked according to prestige awarded inside the community (with the discipline of physics and individuals in receipt of Nobel laureates at the apex). For Hagstrom, the scientific community is like any other functioning social institution in that we can scrutinize it from the outside to discover its structure, function and values. Significantly, with its shared ethos and shared values, his picture of the scientific community is devoutly essentialist. For him, the scientific community has an essence that is composed of shared norms and values.

The second influential conceptualization of the scientific community comes from the work of Thomas Kuhn. *The Structure of Scientific Revolutions*, first published in 1962, was a historical analysis of the construction and articulation of scientific theories (see chapters 3 and 4). The concept of 'paradigm' is central to Kuhn's thesis of change through scientific revolution, and central to the concept of paradigm is the scientific community: for Kuhn, it is through the sharing of a paradigm that we can identify a scientific community. In the first edition of *The Structure of Scientific Revolutions* Kuhn offers little by way of definition of scientific community (Jacobs 1987). Despite Kuhn's relativism, and his recognition of difference between different scientific disciplines, it looks as if he is talking about a general scientific community in the same way as Hagstrom: 'Like the choice between competing political institutions, that between competing paradigms proves to be a choice between incompatible modes of community life' (Kuhn [1962] 1970: 93). This loose form of definition by analogy can be seen as being, at least in part, responsible for generating an understanding of paradigms as being outlooks shared by very large groups of practitioners. Kuhn subsequently revised and refined his understanding of scientific communities in the *Postscript* to *The Structure of Scientific Revolutions*, included in the 1970 edition, and in an essay published in 1977:

> A scientific community consists, in this view, of the practitioners of a scientific specialty. Bound together by common elements in their education and apprenticeship, they see themselves and are seen by others as the men responsible for the pursuit of a set of shared goals, including the training of their successors. Such communities are characterized by the relative fullness of communication within the group and by the relative unanimity of the group's judgment in professional matters. (Kuhn 1977: 296)

This is quite different from Hagstrom's understanding: for Kuhn a scientific community is a small group of practitioners, possibly as small as a few hundred members, who share a particular specialist area of research (there is an obvious parallel here to Fleck's conception of esoteric thought communities). Such a scientific community can be investigated through analysis of the works that they ascribe to, i.e. by analysis of the works that they cite in their publications. However, as well as a scientific community sharing key texts it is interesting that Kuhn's definition includes reference to the shared values of such a community, and he does note the relevance of the work of Hagstrom in the second edition of *Structure* (Kuhn 1970: 176 n.5).

Both Kuhn and Hagstrom are offering, albeit in different ways, an essentialist account of scientific community. The scientific community that they identify has an existence above and beyond that of the actions of its individual members, and it has a core set of values and orientations towards its actions. These two sociological analyses of scientific communities, which represent the standard and the relativist perspectives, both construct an essentialist picture of scientific communities and, by extension, scientists.

Summing up, our media usages correspond broadly to two key sociology of science definitions of scientific community. On the surface these sociological definitions appear to be in conflict with each other: we can either have a general scientific community that comprises all scientists working in all disciplines, or we can see scientific communities as being small, self-contained entities that are bounded by shared specialist subject matter, but are independent of each other. However, many scientists consider themselves to be both members of specialist, small-scale scientific communities that are centred on their specialist research, and also members of a larger community that comprises all those involved in scientific endeavours.

Both major sociological conceptions of the scientific community have received significant criticism in recent years. Hagstrom's account, relying as it does on scientists sharing a value-system that is normative and monolithic, has been attacked for its lack of consideration of variations inside the different specialities making up the scientific community (Crane 1972). Kuhn's account has received more sustained assault, particularly from ethnomethodology and social constructionism. In particular, the work of Karin Knorr-Cetina has been used to show the flaws in both Kuhn's sociological understanding of scientists' organization (Knorr-Cetina 1981) – the community that Kuhn describes is not the group of individuals who are instrumental in producing scientific knowledge – and his understanding of change in science as a whole, in that without a scientific community being properly identified as sharing a paradigm there can be no paradigm shift taking place (Jacobs and Mooney 1997). More generally Steve Fuller identifies the very concept of 'scientific community' as being an artificial construct that emerged in the late 1950s and served a political purpose:

> It would seem that Kuhn managed to 'discover' that science flourishes in self-governing communities just at the time that the democratic instincts of American politicians and political commentators insisted on greater public accountability from scientists. As in so many other cases, a sense of 'community' emerges among disparate individuals as they face a common foe. To put the point in boldest relief, the construction of *science* as *social* served much the same purpose as earlier constructions of the *scientist* as *individual* – only now reflecting the potentially more widespread societal opposition that scientists faced. In both cases, the uniqueness of science is highlighted as demanding special treatment. (Fuller 2000: 209)

Fuller goes on to note that the construction of scientific community that Kuhn put forward in *Structure of Scientific Revolutions* was so stilted and idealized in the way that it minimized observation of any form of disagreement that it deserves to be described as being 'oversocialized'. For Fuller, given this tarnished genesis, the use of the concept of scientific community is not helpful for sociologists of science. His critique can also be applied to the picture of the scientific community that Hagstrom presented. Hagstrom's identification of the values of science embodied in the scientific community produced a landscape that was pretty flat, in that almost all of his respondents conformed to the universally held ideals, and all expressed the commonality of values that he ascribed to the scientific community (broadly similar to those that Merton identified).

As well as these critiques, a shift in method and methodology inside STS has pushed the concept of scientific community even further into the shadows. Where Hagstrom uses a unitary framework to describe *the* scientific community, albeit one made up of the many disciplines inside it, and Kuhn finds communities cohering around a shared paradigm, ANT does something quite different when looking at the people who make scientific knowledge. ANT differs from the two other perspectives on scientific communities in two significant ways. The first is that it isn't a deductive approach: ANT does not try to fit a theoretical framework onto its objects of inquiry in the way that Kuhn or Hagstrom does, although it does impose a conceptual and metaphorical framework, as we shall see. The second is that ANT does not look at 'communities' – discrete and bounded entities that require members to become a part of them actively. Instead, it looks at networks, often loose associations of ties and alliances that are formed pragmatically by actors who have specific goals; very often this goal is the production of scientific knowledge, although ANT can, according to its adherents, be applied to pretty much any instance of social action. The most famous example of an actor network is, perhaps, the first identified using this approach. Michel Callon's study of scallop fishing in Saint-Brieuc Bay, France (Callon 1986), is an approach that is similar in orientation to Bruno Latour and Steve Woolgar's study of a biochemistry laboratory (Latour and Woolgar 1979). It starts with looking at a specific problem/ issue: can the production of scallops be improved through controlling their cultivation? There is, from a scientistic point of view, an obvious way of discovering the answer to this question: set up an experiment and assess the results. However, simply looking at the efforts of marine biologists carrying out experiments would be, in ANT parlance, an instance of 'black boxing': we could see the inputs to the black box in the form of funds and agents, and see the outputs in the form of scientific papers, but the actual processes taking place in the experiment would be hidden from us. What Callon did, and what ANT does, is to challenge the fundamental assumptions of much of sociology of science by looking instead at the controversy that has brought a range of different actors together. In the case of the scallops of Saint-Brieuc Bay, Callon identified four different actors: the fishermen of the bay, who have a goal of making as much profit from their endeavours as possible; the three scientists who are investigating scallop production and who have some very specific expertise and knowledge of scallop farming; the scientists' scientific colleagues, who are not specialists in scallop production but who have an interest in advancing knowledge; and the scallops themselves, which have the goals of proliferating and surviving attacks from predatory starfish. Callon's study examines how it is that the three scientific researchers manage to prevail to the point of conducting their experiment, generating scientific knowledge that is acceptable to their scientist colleagues, and finally convincing the fishermen that their theory – that scallop larvae anchored to a collector will improve yields – is a good one and worthy of their attention. Callon's study follows the course of events taking place and proceeds through describing what is occurring and subsequently identifying and naming the actors involved. But where a more 'traditional' approach in sociology may consider identities of agents to be largely fixed, the ANT approach instead looks at identity work in action, and looks at network formation in action. Actors are enrolled into positions in networks due to some quality they have, or some agency that they are expressing. But these networks are not firmly fixed; they shift and change across time as actors change their position and even their identities. Alliances and allegiances are formed and broken as actors try to achieve their aims. ANT's claim to success is that it tracks these shifts and changes without imposing a strong set of categories on top of actors it has identified.

> Not only was the identity of the scallops or the fishermen and the representatives of their intermediaries or spokesmen (anchored larvae, professional delegates, and so on) allowed to fluctuate but also the unpredictable relationship between these different entities were also allowed to take their course. This was possible because no a priori category or relationship was used in the account. (Callon 1986: 208)

ANT is a very good way of following through events that lead to an outcome, and has clear applicability to how formal science is done (Latour 1987, 2005). Who becomes involved in a process of experimentation and why? How are humans and non-humans enrolled into a network? How do different nodes in the network communicate with one another; how do they deal with controversy? Who is inside and who is outside the network? And ANT clearly sidesteps the whole topic of 'scientific community': if we are enrolled into and disenrolled from different networks, and the actors and actants in those networks have similar status, then the idea of a discrete 'community' that requires some kind of membership ceases to be relevant. However, there are two important problems here.

The first is that wider issues of context become lost in a thin description of interactions. These are, of course, important for any sociological perspective, but must be seen in a wider context. ANT provides powerful tools for looking inside, but negligible cognisance of what the network is located within and why events are taking place. ANT can reply that events occur because actors and actants will them to occur, but we know that the wider socio-economic and cultural contexts surrounding events are significant. Even in the case of the scallops of Saint-Brieuc Bay, the economic and the cultural – who can pay for the product and the reason why the product is consumed – are both completely outside the purview of the ANT approach. The second is a criticism that many post-structuralists laid at the door of the modernists: the replacement of actors' intentions with other things. In ANT the laboratory worker thinks they are analysing data, but actually they are constructing a network of relations with human and non-human actors; the fisherman thinks he is scooping up scallops, but actually he is contributing to the network that is making the scientific knowledge for the marine biologists' colleagues. What does it mean to be identified as being a part of a network when one is not aware of it? Is the network 'real'? What would that mean, in terms of how we might experience it? And have we as sociologists learned anything by assigning or identifying people as being members of networks if they themselves don't think that they are? Why is it that only the ANT theorist can identify the network; what special tools do they have to do this? (See Erickson 2012.)

Scientists construct the scientific community

The general shift towards an SSK frame of analysis killed off the concept of 'the scientific community' as a category for analysis, and it is notable that few studies specifically dedicated to examining the scientific community have appeared in recent years. This is not to say that the concept has disappeared: far from it. A quick search through the *Social Sciences Citations Index* will find a huge number of mentions of 'scientific community'. However, such mentions are precisely that: the scientific community is seen as being an extant institution that needs no further analysis and is deployed in these journal articles without question.

We could leave the matter there, simply noting that there is a common-sense perception of the scientific community (as seen in, for example, contemporary media representations) which sociology of science and STS can debunk or dismiss quite easily, were it not for one further complicating factor. The problem here is that descriptions of a united scientific community are not restricted to the media and the general public. If they were we could explain their appearance as being due to, perhaps, a lack of knowledge of how scientific activity is structured. The problem is that formal scientists, too, present themselves in public and in private in this way, and a charge of lack of knowledge will not stick here. Here are some examples of scientists talking to the media, all culled from broadsheet daily the *Guardian*'s online archive.

[1] We would argue that since scientists helped to create nuclear weapons, the scientific community today has a profound responsibility to help reduce and ultimately disarm them. (Martin Rees and Des Browne, 'Science's nuclear responsibility. Scientists have a critical role to play in reducing and finally eliminating nuclear weapons', *Guardian*, Sunday 4 April 2010)

[2] Some things worry me though. Few in the science community in the UK would agree with EU thinking on GM crops, and there are concerns about the EU's position on data protection and health research for example. However, I think that influencing from within would be the most effective way forward. Scientific evidence will more often than not carry the day, as can be seen from the EU's leadership on climate change.

(Paul Nurse, 'Science funding and the EU: you've got to be in it to win it. The benefits to UK research, from finance to international collaboration, make a strong case for continued EU membership', *Observer*, Sunday 27 January 2013)

[3] BC: I think one of the great challenges for the scientific community is how to deal with arguments from people with genuinely held views that are demonstrably wrong and potentially damaging. I'm thinking of issues like the vaccination of children or the imperative to reduce greenhouse gas emissions. The science is very clear on these issues, and science really is the best guide we have to facing global challenges. The dilemma is how to convince quite vocal minorities that a rational and scientific approach is no threat to their political or religious beliefs – it's just the best approach. (Stephen Hawking and Brian Cox, 'Gods of science: Stephen Hawking and Brian Cox discuss mind over matter. We paired up Britain's most celebrated scientists to chat about the big issues: the unity of life, ethics, energy, Handel – and the joy of riding a snowmobile', *Guardian*, Saturday 11 September 2010)

What do these designations of scientific community mean? As in our deconstruction of the journalists' usage of the term 'scientist' earlier, we need to use a certain amount of inference to identify who is being referred to here. It would appear likely that in example (1) the 'scientific community' being referred to is that of physicists because a quite specific object is being referred to (nuclear weapons). However, it could be that Martin Rees and Des Browne do indeed mean 'all scientists': these writers may consider it the responsibility of all scientists to play a role in the reduction and ultimate abolition of nuclear weapons. This is not necessarily far-fetched: people involved in scientific research are often members, or adherents, of collective organizations, such as Scientists for Global Responsibility (www.sgr.org.uk), that do make representations on behalf of all of their members to promote collectively held values.

Example (2) also uses the term 'scientific community' in an ambiguous way: many formal scientists are involved in genetic modification of crops, but the vast majority are not and may have little knowledge of what is, after all, a specialized area of formal science. In this example I think that we can infer that the scientific community being referred to is a discipline-specific one; geneticists. But we should also note that Sir Paul Nurse is writing this piece in his capacity as president of the Royal Society, a very eminent and influential organization that, as a 'fellowship of the world's most eminent scientists and . . . the oldest scientific academy in continuous existence' (Royal Society website, https://royalsociety.org), acts as a voice for all of formal science.

Finally, example (3) is unambiguously collecting all formal scientists into a single community. It sets science, the scientific method and this very large collective (the scientific community) against a significant 'other' – anti-science, foolishness and wrong-headed thinking.

Up to this point we have seen three different accounts of scientific community emerging. The first, the common-sense notion of a community of scientists, is seen clearly in media productions, particularly news media. The second, the sociological account of scientific community, is largely absent from contemporary sociology of science, but has a long tradition of identifying a community that is structured around shared values and goals. The third account, that of scientists themselves, is seen in their self-descriptions. Here elements of the 'common-sense' definition of the scientific community are mingled with personal experiences of working in collective ways to produce accounts that are personalized, yet reflective of wider societal understandings of what science is. All three accounts promote an essentialist picture of science and scientists, but construct the idea of 'scientific community' in quite different ways.

We need to remind ourselves that individual and collective identities are being constructed in relation to each other and in the context of a constructed culture. In this specific case, scientists construct their identity with reference to their membership of a collective of scientists, and the result is they feel that they are a part of 'the scientific community'. But, as Fleck ([1935] 1979) and a great many other sociologists have reminded us over the years, scientists do not exist in, or emerge from, a vacuum even though many cultural representations make it look as if they do. Here we must recognize that scientists' construction of collective identity is made not only by reference to other scientists, but also by reference to wider social and cultural representations of science, scientists and scientific communities. Most scientists can define themselves in relation to the collective of which they feel they are a part, and are doing this defining in a cultural context that includes clear reference to their collective, and even has a clear construction of an archetype of identity for the members of that collective. Scientists, as an occupational group, are not unique in this respect, but are part of a very small minority in our society. Whilst we can find stereotypical descriptions of a range of distinct social groups, often pejorative ones, it is rare to find social constructions of occupational classifications as a whole. Journalists, politicians and estate agents spring to mind as being the victims of unfair negative stereotypes, but there are few others. We wouldn't, for example, find similar constructions concerning engineers, nurses, factory workers or call-centre operatives. It could be argued that one reason for this is the morally contestable aspect of the work carried out by some scientists (Bradley et al. 2000: 180). Many scientists are seen by the public as being involved in work that challenges their moral and ethical standpoint, or even threatens them personally. High-profile media coverage of contentious scientific research such as human cloning or genetic modification of organisms adds to these perceptions. And we can see that some representatives of the scientific community feel obliged to fight back against these perceptions of science, often very vocally. We can certainly feel sympathy for scientists whose hard work and skilfully produced knowledge is ignored or demeaned in public; witness climate change deniers being given high-profile coverage in the media.

However, other aspects of scientific communities and institutions, notably their systematic discrimination against women and people from minority ethnic backgrounds, receives much less coverage and does not deserve our sympathy. Zygmunt Bauman (2001) notes that whilst community is always seen as a good thing, its inclusive qualities are meaningful only with reference to what it excludes. What is a place of security for those who 'belong' can be an alien and unfriendly environment for those who do not conform or identify with the majority or the mainstream within it (Erickson et al. 2009: 234). Self-defining scientific communities provide a good example of groups that systematically exclude and discriminate, notably against women. This is plainly visible in recent analyses of scientific careers.

A career in formal science?

We've already seen that definitions of the scientific community are diverse and contested. In this next section we will focus on studies of academic researchers; this is a pragmatic move in that quantifying who is a member of a scientific community requires some kind of definitional device. In this case, the definition is largely congruent with my category of 'formal science', i.e. professional academic researchers involved in producing scientific knowledge. A 2012 report from the

Equality Challenge Unit uses Higher Education Statistical Agency (HESA) data to construct a picture of the UK HE workforce, and the focus here will be on SET staff. HESA data identified 97,065 SET academic staff in UK HE institutions in 2012 (Equality Challenge Unit 2012: 22).

In SET departments, 60 per cent of academic staff were men, and in some departments the gender imbalance was particularly striking. For example, electrical engineering and computer engineering had 13.8 per cent women academic staff, mechanical and aeronautical engineering 16 per cent and physics 16.8 per cent (Equality Challenge Unit 2012: 45). Men made up the majority of academic roles across all modes of employment, SET and non-SET. The gender difference is most noticeable among academic senior managers in SET departments, where 77.0 per cent of staff were men. Among SET professors, 15.6 per cent are female, in contrast to 25.8 per cent of non-SET professors. This disparity in terms of women in senior posts contributes to a 13.7 per cent median gender pay gap between men and women for academic professionals across the whole UK HE sector.

Black and minority ethnic (BME) academic staff are also under-represented in higher academic and managerial posts, although a higher proportion of professors (6.8 per cent) were BME in SET than in non-SET departments (4.8 per cent). A higher proportion of white than of BME academics earned over £50,000. In every mode of employment, a higher proportion of UK BME academic staff were on short-term contracts than of white staff. This contributes to a 5.6 per cent median pay gap between white and BME academic professionals across the whole UK HE sector.

The majority of disabled academic staff work in non-SET departments; 2.3 per cent of academic staff in SET departments were disabled. A smaller proportion of SET professors were disabled (1.9 per cent) than of non-SET professors (2.4 per cent). However, the data found that there was a negative pay gap between non-disabled and disabled academic staff (Equality Challenge Unit 2012: 147).

Clearly there are significant differences between groups of staff in the UK scientific community, with women, disabled and BME staff consistently under-represented in senior positions in all departments.

The UK figures are similar to findings in other countries. For example, the US National Science Foundation found that 22.3 per cent of full-time full professors with science, engineering or health doctorates in 2010 were women, and the share of full-time, full professorships held by under-represented minorities is lower and has risen more slowly than the share held by women (National Science Foundation 2013: 9). Female scientists in the United States earn much less than men, on average, with the 'differences varying strongly by field' (Shen 2013: 24):

Median salaries:
Biology:
    Men $65,000
    Women $50,000
Chemistry:
    Men $79,000
    Women $62,000
Physics and Astronomy:
    Men $89,000
    Women $54,000
18% average pay gap, all positions. (ibid.)

Shen identifies the 'leaky pipeline' structure of women's careers (see next section). 'In biology, for example, women comprise 36% of assistant professors and only 27% of tenure candidates' (ibid.: 22).

In France, the Centre national de la recherche scientifique (CNRS) plays a major role in public research, running approximately 1,300 laboratories and employing 11,626 researchers (de Cheveigné 2009: 114).

> Compared with their proportion in the general French population women are under-represented among French CNRS researchers. . . . 31.2% [were women]. These proportions have changed extremely slowly. Indeed, when the CNRS was created in 1946, 30% of the researchers were women. . . . [W]omen also have trouble moving up the hierarchical ladder. This 'glass ceiling' effect is a kind of invisible barrier to women's career development, and indeed it has to be invisible because French law prohibits discrimination between men and women. (ibid.: 115)

Further afield, similar patterns of disadvantage for women in the formal scientific community can be found. UNESCO's extensive report looking at women in science around the world found that: 'While the total number of women full professors increased by 23 per cent from 1999 to 2003 (compared to a 13 per cent increase in the number of men), the proportion of female professors in 2003 was still only 15 per cent, compared to 13 per cent four years earlier' (UNESCO 2007: 76). Women constitute only slightly more than one-quarter of the world's researchers (ibid.: 117) although there are some notable exceptions; in Latin America and the Caribbean, 43 per cent of researchers are women (ibid. 118). UNESCO concludes that 'There remains an absence of researchers and women scientists in top managerial positions throughout the world. There are a wide range of factors that may explain the lower number of women in senior research and development, including work–life balance, gendered patterns and approaches to productivity, and performance measurement and promotion criteria' (ibid.: 12).

Gender discrimination in science

Why is it that women, in particular, are being prevented from achieving senior academic positions and visibility in the scientific community? There is a range of traditional explanations, including ludicrous and objectionable biological arguments suggesting that women are inferior to men, and these indicate an institutional sexism in the scientific community. However, whilst sexism may be at the root, we need to take a more detailed look at the career structure and operation of formal science to understand the specific difficulties women face in advancing their careers. Needless to say, difficulties take a variety of forms.

The structure of women's careers as a whole takes the form of a 'leaky pipeline'; many women enter formal science careers, but few achieve the highest levels, either dropping out along the way, or finding their careers impeded by a range of factors. This starts early in women's careers:

> The proportion of women PhD students planning a career as a research chemist falls from 72% in the first year to 37% in the third year because of factors which become apparent to them during their PhD study. . . . Overall only 12% of women PhD students in their third year plan to remain in academia compared to 21% of men. Of those women who do plan to remain in academia, on average they plan to remain for less time than men in the same position. (Newsome 2008: 1)

What were these factors that become apparent during these women's PhDs? An obvious one is that the standard academic science career requires a lot of moving around and is in many ways inimical to having a stable family life. Interestingly, in interviews women PhD chemistry students said that their image of academic chemistry as 'anti-family' came not from male academics, but from other female academics labelling 'motherhood' a dirty word. For example:

> 'My supervisor doesn't understand why anybody would want children, why anybody would even contemplate doing anything except researching. She'll come out with snide comments like "You're not pregnant are you?"'. . . . 'I had a friend who got pregnant during her PhD and the female supervisor just said, "well that's the end of your PhD, see you later, you can come back in six months time but the chances of getting the productivity we should out of you are probably quite slim so . . ."' (focus group 3 participants, second year women). (Newsome 2008: 37)

Female PhD students also reported that they were uncomfortable with the 'macho' culture of their research group, felt isolated and excluded – and in some cases bullied – and formed the impression that doctoral research is an 'ordeal filled with frustration, pressure and stress' (ibid.: 7). The report concludes that the chemistry PhD programme and academic careers are modelled on masculine ways of thinking and doing, 'which leaves women neither supported as PhD students nor enthused to remain in research in the longer term' (ibid.).

Lack of visibility of feminine role models was also widely reported in Newsome's research. 'Female academics were described by female [research] participants as extremely ambitious, competitive and aggressive. It was the understanding of female participants that these women have had to be this way, in order to achieve a senior position in a "testosterone-fuelled" environment' (Newsome 2008: 37):

> 'You look at lecturers in chemistry and out of about 30 people, there's two women. From statistics you think, "It's going be quite hard to be one of those two women". Plus the women who do succeed are very strong minded, really strong women that have to fight their whole way up and I don't know if I could do that.' (Sue, completed). (ibid.)

Overall, many women chemistry PhD students simply come to the conclusion that academic careers are too all-consuming, too solitary and not sufficiently collaborative, and this is compounded by the negative advice they get concerning future challenges and sacrifices.

There's no reason to think that other formal science PhD programmes provide a more supportive environment for women students, and given this it is not surprising that the pipeline starts leaking at this point. Those women that do pass through this point face the challenge in their thirties of making a difficult choice between family and career. De Cheveigné's interviews with seventeen French researchers found that '[e]very female researcher we questioned acknowledged the difficulty of leading a family life just at the time they needed to lay the foundations of their future career. The difficult period is unanimously situated between the ages of 30 and 40 years old' (de Cheveigné 2009: 127). Despite claiming that they were not actively discriminated against, de Cheveigné's research participants could clearly identify unfair treatment from senior colleagues. For example:

> 'I asked to be 80% part-time for a year when my second child was born, my director replied that I could no longer be at the top on my research subject! And yet, I work alone on my subject and I wouldn't have jeopardized my research because I am still as enthusiastic about it and afterwards it's a matter of organization: I can also work well at home. In an intelligent structure like the CNRS, it's nonetheless surprising to meet people like that, attached to macho principles but anyway, that's just one person and among the younger generation it doesn't happen like this. (Female 32 y.o. early career researcher). (de Cheveigné 2009: 128)

It is not difficult to think of institutional support mechanisms, such as supported maternity leave and protected research time on return to work, that could address this issue. Indeed, de Cheveigné's conclusion is that there are a number of mechanisms that penalize women more than men within 'a research system which, through poor internal communication and weak human resource management, does not sufficiently support its personnel. Most fundamentally, women tend to mobilize a more collective model of scientific research than do men, but men fare better when the evaluation procedures focus on the individual' (de Cheveigné 2009: 132).

Institutional procedures and mechanisms can, and should, be addressed through introduction and enforcement of equal opportunities policies. More difficult to tackle are informal and unseen mechanisms that serve to exclude women from promotion or from being more visible in the scientific community.

In addressing the issue of visibility in one discipline – evolutionary biology – Schroeder and colleagues looked at the ratio of women to male presenters at a major conference, the European Society for Evolutionary Biology (ESEB) Congress 2011. Academic science conferences present a range of events and materials: academics present posters and papers describing their recent work, but conferences usually feature invited speakers to present their cutting-edge research as indicative of where the discipline is going. Schroeder et al.'s analysis of the gender of invited speakers revealed a significant point: 'Although 23% of all initially invited speakers . . . were women, only 15% of the realized invited speakers were women. This reduction was because 50% of invited women declined talks compared to 26% of invited men' (Schroeder et al. 2013: 2065).

Schroeder et al. considered three, possibly interrelated, mechanisms that could explain this. The first – that there was a larger pool of male than female evolutionary biologists to draw from – could be discounted immediately, as roughly the same number of men and women were giving ordinary paper and poster presentations. The second was already known: more women than men declined invitations and this was most likely due to women having child or other care responsibilities that precluded them from travelling. The third could be implicit bias, which is:

[a] known cause for women to be at a disadvantage when climbing the career ladder. Both males and females subconsciously treat and perceive women and men differently, even if they are equally skilled and experienced. People tend to assign fame more often to men than women. Seeing mainly male invited speakers may reinforce an expectation that matches 'invited speaker' with 'male', leading to fewer women being invited. (ibid.: 2068)

The overall result, at least for this part of the scientific community, was that participants at the conference were exposed to fewer women presenting their excellent research, resulting in a feedback loop which reinforces the idea that it is mainly men who do the best research.

There are further structural issues that disadvantage women. The UNESCO report found that whilst women are as likely as men to collaborate on research projects, they tend to belong to smaller teams, so that their rate of return on collaborations is less than for men. In addition, women co-author less than men, and men publish a higher number of shorter papers from their research (UNESCO 2007: 125).

This conclusion is corroborated by more recent bibliometric research. Larivière et al.'s study of global formal science research outputs in the form of academic publications found that men dominate scientific production in almost every country. Many scientific papers are multi-authored and, globally, women account for 30 per cent of fractionalized authorships. 'Women are similarly underrepresented when it comes to first authorships. For every article with a female first author, there are nearly two (1.93) articles first-authored by men' (Larivière et al. 2013: 212). Larivière et al. point out that this is a poor strategy on the part of scientific institutions and the scientific community. Whilst the private sector has long recognized the business case against unfair discrimination (Becker 1971; Erickson et al. 2009: 205–6), particularly in that it ignores important talent that can lead to better productivity, science is lagging far behind: 'For a country to be scientifically competitive, it needs to maximize its human intellectual capital. Our data suggest that, because collaboration is one of the main drivers of research output and scientific impact, programmes fostering international collaboration for female researchers might help to level the playing field' (Larivière et al. 2013: 213). This kind of exclusion is, like the lack of female invited speakers at a conference, difficult to notice or diagnose as there is such a persistent culture of patriarchy inside the scientific community. Yet if we don't attend to the 'unseen' barriers we risk reproducing the prevailing conditions. Liisa Husu's research shows that there are what she calls 'hidden roadblocks' facing women researchers:

> [I]t is not only the things that happen to women – such as recruitment discrimination or belittling remarks – that affect them in pursuing a career in science or that slow their career development. It is also the things that do not happen: what I call 'non-events'. Non-events are about not being seen, heard, supported, encouraged, taken into account, validated, invited, included, welcomed, greeted or simply asked along. They are a powerful way to subtly discourage, sideline or exclude women from science. A single non-event – for example, failing to cite a relevant report from a female colleague – might seem almost harmless. But the

accumulation of such slights over time can have a deep impact. (Husu in Al-Gazali et al. 2013: 38)

Box 5.2 Gender discrimination in society

The vast majority of gender discrimination in UK society is against women. It takes a range of forms, from blatant prejudice, such as refusing to appoint women, to more subtle forms, such as undermining women in the workplace through assigning them stereotypical or demeaning roles. Frequently, gender discrimination is attached to women's reproductive roles.

Despite forty years of anti-discrimination legislation – itself a result of concerted action by feminists – gender inequality persists, particularly in the world of work. Pay gaps may have narrowed, but still exist, and women in the UK labour market are clustered in low-paid, low-status jobs, excluded from positions of power. Surveys continue to show that women carry out about 70 per cent of domestic labour, with men cherry-picking the tasks they like to 'help' with (Bradley 2013: 132).

The ideology of gender is deeply entrenched in British culture, reproducing gender stereotypes and facilitating patriarchal attitudes. A frequent explanation for gender stereotypes is that they are reflections of the 'naturalness' of male and female social roles. The relationship between the ideology of a 'natural' gender order and the workplace is mutually reinforcing: people make choices about their paid and unpaid work on the basis of such ideologies, while at the same time the world of work mirrors and reproduces gender stereotypes (Erickson et al. 2009: 219).

Gender discrimination is endemic in society and rife in formal science. This is, sadly, evidence for the thesis that science is deeply and intimately connected to society; we live in a sexist society, so we should not be surprised that formal science is also sexist. Yet it is not unreasonable to expect better of the institutions of formal science given their relentless promotion of their adopted values of objectivity and universalism. There is nothing objective and only particular sectional interests in gender, and other forms of, discrimination.

Sexism

Running parallel to, and sometimes through, these structural barriers to women's progression in the scientific community are sexist attitudes, language and behaviour. In interviews I carried out for a paper on scientific vocation (Erickson 2002), I found that at job interviews almost every single female formal science researcher had been asked, or could recognize the implicit question of, whether they intended to take a career break to have children; none of the men interviewed reported the same. Such questions are, of course, illegal under UK employment law, but women researchers recognized that they had to put up with this to keep their careers on track. Harassment and bullying, particularly by male senior colleagues towards women junior colleagues, is well recognized in the scientific community, although not so often acted upon. The journal *Nature* took it upon itself to raise this issue in an editorial, responding to the large number of women who have come forward to describe their own experiences of misogyny and harassment in the workplace. *Nature* notes the corrosive effect of sexual harassment in the workplace, damaging people, blighting careers and reducing productivity.

> A major problem is the widespread tacit acceptance of adolescent behaviour. Let us call him Dr Inappropriate: he is the lecturer at the conference drinks reception with the wandering hands. . . . He is the head of department who thanks his female colleague for her excellent presentation but suggests that she wears a shorter skirt next time (yes, this really happened). Worse, Dr Inappropriate is often the lab head, or an equivalent – a mentor with responsibility and power over the careers of the women whom he asks to work late on a project or join him in a taxi home. Sometimes he is a very senior scientist indeed. . . . The evidence of the scale of the problem is anecdotal [but] just ask around: everyone knows a Dr Inappropriate. (*Nature* editorial, 24 October 2013: 409)

Their call for anti-bullying and anti-harassment policies to be rigorously enforced, and, perhaps more importantly, for other scientists to challenge Dr Inappropriate, is welcome. But the fact that such a statement needs to be made in the leading international journal of science in 2013 is in itself quite shocking. Dr Inappropriate's behaviour, and the culture of complacency and inaction surrounding it, would not be tolerated in any other UK public or private institution.

This is a major problem for the scientific community, and it would appear to be very widespread. After a friend revealed that she had been sexually assaulted by a colleague at a field site run by a university, biological anthropologist Kathryn Clancy decided to investigate further. Clancy and three colleagues set up an online survey to explore biological anthropologists' experiences of sexual harassment in the field. They found that abuse was widespread (70 per cent of women reported sexual harassment, as opposed to 40 per cent of men), systemic and played out along lines of power. Young female graduate students were usually the targets; older, more senior men were usually the perpetrators.

Women were significantly more likely to have experienced sexual assault: 26% of women vs. 6% of men in our sample [of 637]. . . . The perpetrators of harassment and assault differed between men and women. Harassment aimed at men primarily originated from peers at the field site (horizontal dynamics) whereas they originated from superiors when directed toward women (vertical dynamics). (Clancy et al. 2014: 4)

Addressing gender discrimination in science

It appears, then, that little has changed since Harriet Zuckerman and Jonathan Cole identified the 'triple penalty' that faced women scientists in the 1970s:

> First, science is culturally defined as an inappropriate career for women; the number of women recruited to science is thereby reduced below the level which would obtain were this definition not prevalent. Second, those women who have surmounted the first barrier and have become scientists, continue to be hampered by the belief that women are less competent than men. Whatever the validity of this belief, it contributes to women's ambivalence towards their work and thereby reduces their motivation and commitment to scientific careers. And third . . . there is some evidence for actual discrimination against women in the scientific community. To the extent that women scientists suffer from these disadvantages, they are victims of one or more components of the triple penalty. (Zuckerman and Cole 1975: 84)

The triple penalty is, finally, being addressed by institutional action, at least in the UK scientific community. Founded in 2005, the Athena SWAN Charter is a recognition scheme for UK research institutions, funded by the Equality Challenge Unit, the Royal Society, the Biochemical Society and the Department of Health.

> Any university or affiliated research institute that is committed to the advancement of the careers of women in STEMM [science, technology, engineering, maths and medicine] can become a member of the Charter, accepting and promoting the six Charter principles:
>
> - to address gender inequalities requires commitment and action from everyone, at all levels of the organisation
> - to tackle the unequal representation of women in science requires changing cultures and attitudes across the organisation
> - the absence of diversity at management and policy-making levels has broad implications, which the organisation will examine
> - the high loss rate of women in science is an urgent concern which the organisation will address
> - the system of short-term contracts has particularly negative consequences for the retention and progression of women in science, which the organisation recognises

- there are both personal and structural obstacles to women making the transition from PhD into a sustainable academic career in science, which require the active consideration of the organisation. (Equality Challenge Unit 2013: 4)

In 2012 there were 179 UK HE institutions that had achieved at least the 'bronze' accreditation level of the Charter. The funders of the majority of formal science research in the UK, represented by Research Councils UK (RCUK), are bringing on stream proposals to ensure that recipients of funding are committed to equality and diversity in research. 'At this time, RCUK does not require formal accreditation, such as Athena SWAN, for grant funding; however, we will be reviewing and may consider such measures if there is no evidence of improvement' (RCUK press release, 'RCUK announces policy to support equality and diversity in research', 17 January 2013).

However, these institutional responses address only the *external* factors in discrimination, such as lack of institutional support, lack of role models, discriminatory practices. What they don't address are the *internal* factors inside scientific communities, and the internalized values and attitudes that form the cultures inside esoteric scientific communities. It is clear from the summary of studies given earlier that esoteric thought communities import extensive sexist and patriarchal discourses and ideas from exoteric communities their members are a part of.

Conclusion

Sociology of science has shifted its focus on the scientific community. In one respect this is a good thing. The essentialist construction of a scientific community that has a life of its own beyond that of its members, and embodies and expresses the values of its members, is not helpful to us in making sense of scientists' work and science in society. Such a construction, clearly visible in the work of Hagstrom and Kuhn, was a useful device for sociologists to characterize a social institution in the context of a form of sociological analysis that looked for systemic understandings of the world. It was also a useful device for scientists themselves, particularly at times when science was under attack or under public scrutiny: finding a self-regulating institution that embodies the core values of the society it was 'serving' is very useful in political terms. However, such understandings, if they were ever accurate, are inappropriate for us today. The scientific community that is visible in contemporary society is certainly not the monolithic scientific community described by Kuhn and Hagstrom, although some elements of their description, such as the idea that specific disciplines will form their own communities through sharing knowledge, are still visible. Formal scientific activity, i.e. where formal scientists are actively working, is now fragmented into sub-disciplines which are small in size and have a diminishing number of connections to other sub-disciplines.

However, that does not mean that we can dispense with the concept of scientific community. Apart from anything else, formal scientists, the institutions they are a part of, and funding agencies clearly consider the scientific community to be a thing in its own right. But sociology of science has shifted away from looking at collectives to look at individuals involved in the production or construction of knowledge: ANT is in a hegemonic position with respect to sociology of science. Such studies are illuminating of the local processes that occur in the research process in specific locations, and shed a great deal of light on the forms of interaction, negotiation and argumentation that are necessary for scientific knowledge to emerge. What they cannot show, indeed what they occlude, are the external and collective factors that are significant in the everyday lives, work and careers of scientists. Using a conception of scientific community similar to that of Fleck could alter this.

Fleck's idea of exoteric and esoteric thought communities, groupings of people characterized by a particular style of thinking, is a useful model for making sense of the multiple scientific communities that exist. Fleck notes that people can be members of many exoteric thought communities (such as political parties, clubs and societies, social communities, families) but only a small number – maybe only one at a time – of esoteric thought communities. He also notes that the exoteric and esoteric will have an impact upon each other: an exclusive and isolated thought community is simply not a possibility, in the same way that Wittgenstein considers a private language to be impossible.

Fleck's concept of thought communities allows us to explain why it is that scientists themselves will identify membership of more than one exoteric community (for example a national identity, a religious community, a sports community, neighbourhood group, trade union, political party and so on), but will also usually identify a strong affinity with a single esoteric community (as a microbial geneticist, or a carbon chemist, etc.). This picture, of a scientific community that is fragmented and ordered around tight specialisms, concurs with the current organization of scientific research, and with the current state of wide-ranging generic scientific bodies that represent 'all scientists' such as the Royal Society.

By looking at the entry of external styles of thought and ideas into esoteric thought communities, Fleck also provides us with, at least, a starting point for understanding the continued discrimination against women in the formal sciences. In addition, his schema allows us to consider how agendas inside formal science structures are set according to external, frequently androcentric, priorities. The exoteric scientific community is a social construction that is made from the actions and meanings of all members of society. These do not have similar impact, and it is clear from the media representations of scientific communities and scientists that there are two powerful discourses that articulate a dominant idea of what the scientific community is: the voices of leading scientists who represent scientific institutions, and the discourse of scientism that impresses upon us the idea that science is external, neutral and the locus of a form of knowledge that is better than others available. But there is a further contributory factor that is important here: the actions and language of the members of esoteric scientific communities. It is here, as we discover when we look inside, that we find significant challenges to the idea of a neutral and objective set of practices that are disinterested. Quite the reverse is visible, with open forms of discrimination, particularly against women scientists, taking place. These practices show in graphic ways that the exoteric thought communities of which scientists are a part – neighbourhoods, political parties, social groups, society in general – have a direct effect on the thought styles of esoteric scientific communities, as the sexism that is a part of these wider groupings comes to be reproduced in the scientific workplace and institution. But this is a two-way street; as we have seen, scientism is a contributory factor to maintaining unfair hierarchies and modes of discrimination. Scientism serves to exclude morals and values from scientific knowledge, which, being the only goal of scientific work, places such questions as having a second order of importance for esoteric science communities. This is not meant to be an accusation of complacency against formal scientists: the vast majority will be appalled at the sexist and discriminatory practices taking place across formal science. But it is to point out that considerations of the value and moral structures of scientific institutions are of a second order to the stark goal of achieving maximum scientific knowledge outputs.

Science is being actively constructed inside scientific communities, but also outside of scientific communities in society and culture. These exoteric constructions of science play a role in shaping what scientists think of their work, their project and themselves. Society also articulates values and morals through discourse, and these are internalized and rearticulated in smaller thought communities. Given that our society and culture are sexist it is no surprise that scientific communities are also sexist. Given the odds stacked against them, and the persistent stereotyped representations in contemporary culture of them, it is remarkable that women scientists have achieved so much to address issues of discrimination and sexism in scientific communities.

Further reading

Robert Merton's ground-breaking work in sociology of science and scientific communities is brought together in his 1973 collection:

Merton, R.K. (1973) *The sociology of science: theoretical and empirical investigations*, Chicago: University of Chicago Press.

Lots of good material on the structure, composition and hierarchy of formal science is available online. Of particular interest to understanding gender discrimination are:

National Science Foundation National Center for Science and Engineering Statistics (2013) *Women, minorities, and persons with disabilities in science and engineering: 2013*, Special Report NSF 13-304, Arlington: National Science Foundation. Available at http://www.nsf.gov/statistics/wmpd

OECD (2006) *Women in scientific careers*. Paris: OECD. Available at http://www.oecd-ilibrary.org/employment/women-in-scientific-careers_9789264025387-en

For an excellent introduction to gender, society and social analysis see:
Bradley, H. (2013) *Gender*, Cambridge: Polity.

6

Popular Science

Opinion among scientists, among whom I am one, will insist that, 'That which we don't understand' means only 'That which we don't yet understand'. Science is still working on the problem. We don't know when or even whether we shall eventually be brought up short.
> Richard Dawkins, *A Devil's Chaplain* (2004)

In previous chapters we have seen how scientists produce science, or explain science, in a formal scientific way. The standard account of this process is summed up clearly by Dawkins in the epigraph to this chapter: science is working and it will ovecome lack of understanding. In this chapter, by contrast, what we see is scientists explaining science in non-formal scientific ways, and, sometimes, promoting science in ways that could be construed as being anti-scientific. Many popular science books promote a version of science that is contrasted starkly against the 'other' of science – or at least the 'other' that popular science writers want to identify – namely foolishness. Yet there is no necessity for one to accept this formulation, i.e. we do not need to accept the hypothesis that we can either be scientific in our outlook or we can be foolish. And yet that formulation is repeatedly made in popular science texts, and in most cases on the basis of almost no evidence. Let's take a quick example to illustrate how popular science articulates a discourse of 'science versus foolishness'.

Sir Paul Nurse is the president of the Royal Society, one of the most prestigious, and oldest, bodies of scientists in the world. He is also a Nobel Prize winner (medicine 2001), so is clearly highly respected in the scientific community and further afield. In January 2011 Nurse was given a whole episode of BBC2's flagship science documentary programme *Horizon* to himself to look at the status of science in contemporary society. The programme was titled 'Science Under Attack'.

In it, Nurse presents the audience with a passionate call for a better understanding of science, and for scientists to present their work in more open and understandable ways. Science, Nurse tells us, with its staunch scepticism, objectivity and endless questioning of itself, can provide answers but only when all evidence is examined correctly and appraised using impartial perspectives. This is very much the line that we find in the standard account and in scientism; it would be unrealistic to expect anything else from a leading scientist.

Then Nurse turns his attention to the controversy surrounding genetically modified (GM) crops, a subject which, he thinks, shows a 'mutual misunderstanding between the scientists and the public'. He meets a leading GM crop scientist and the programme shows footage of anti-GM campaigners ripping up crops and denouncing GM foods as bad for the environment and for people. Nurse goes on (this is a verbatim transcript, commencing at 53'25"):

> The controversy surrounding GM was something I really wanted to understand. I went and talked to members of the public to find out why they were so against it. And one thing that came up very often was that they were against eating food with genes in it, and that's something that would never occur to a scientist, as a scientist obviously knows that all food has genes in it, but, I mean, why should a member of the public know that? What had happened here is that we scientists hadn't gone out there and asked what bothered the public, we hadn't talked to them about the issue, not had dialogue with them.

This is quite an astonishing statement from one of the most prominent and respected scientists in the UK. Nurse has spoken to 'members of the public'. Who, when, how? What did he ask them? What does he mean by 'very often'? I would not let my undergraduate social research methods students get away with this level of vagueness. Personally, I have never met a single person who does not want to eat GM foods because the food might have genes in it. Does Nurse really think that all those anti-GM campaigners (that he actually filmed) don't understand what genes are, that they are against GM because they might eat food with genes in it? Does he really think that the general public are so ignorant of science and that only scientists know that 'all food has genes in it'? This idea that people don't want genes in their food is not a scientific proposition; not even a sensible proposition. But bear in mind that earlier in the programme Nurse had pointed out that science is so good because of its staunch objectivity, its impartial methods, its constant scepticism. Nurse clearly doesn't consider those things apply to him in this piece of 'research'. Feyerabend observed in *Science in a Free Society* (1978b) that scientists are often unscientific about other knowledge claims, and rely on assertion and status to critique other forms of knowledge.

Secondly, Nurse is very dedicated to science, as we might expect, but seems to be completely unaware that social scientists have carried out extensive research into why the public do not want GM foods, much of which shows that opposition centres on a critique of multinational corporations and a concern that GM is 'against nature'. Nurse's refusal to engage with social science research implies that it is of lesser status than formal science.

Thirdly, Nurse has, from his 'research', found out why GM foods have not been adopted in the UK. It is because the general public are foolish and they have not understood the scientific arguments concerning GM. It is not only the public who are to blame: scientists are culpable too. If only scientists had explained their case to the public better then GM foods would have been adopted in the UK. Again, this is very far away from being a scientific proposition, but is a clear statement of value and attitude from Nurse.

These themes – the credulity of the lay public, the importance of communicating science to educate the lay public, the superiority of scientific knowledge to other kinds of knowledge – are recurrent themes throughout the genre of popular science. Popular science will often place science in opposition to its 'other' and will, either directly or indirectly, identify the 'other' as being ignorance, foolishness, credulity and/or faith.

What is popular science?

Popular science, particularly in book form, was in vogue throughout the twentieth century, and remains common in contemporary culture. Inside the genre we can distinguish texts that seek to popularize science for a lay public and those that seek to 'save' the lay public from the dangers of pseudoscience, quackery and false scientific analysis. Whilst both forms of texts achieve a similar goal of reinforcing the position of pre-eminence that science holds in terms of the status of knowledge in society, they achieve this in different ways. The genre of popular science books that specifically discuss the gullibility of the general public and seek to preserve them from 'false' science only fully emerged in the 1950s, but is still a strong trend in today's popular science.

The roots of this genre lie in the nineteenth century with writers such as Thomas Huxley, who wrote popular accounts and defences of Darwin's theory of evolution, and was involved in significant public debates in the 1860s and 1870s, and John Tyndall, an eminent physicist. Both men had work published by the newspapers of the day and in book form. The tradition of explaining scientific subjects to non-scientific publics begins in the early nineteenth century with the public lectures presented at the Royal Institution in London, a tradition that continues to this day with the Royal Institution annual Christmas lecture, given by an eminent formal scientist to an audience of young people.

The twentieth century saw a proliferation of texts popularizing and explaining science to the lay public. The authors were often eminent scientists themselves – for example, James Watson wrote his idiosyncratic account of his discovery, with Francis Crick and Rosalind Franklin, of DNA in 1953 (Watson 1968); Albert Einstein wrote a popular account of his theory of relativity that ran to fifteen editions between 1916 and 1952 (Einstein [1916] 2001). The emerging genre of science fiction (see chapter 7) provided another vehicle for explaining science; authors such as H.G. Wells saw themselves first and foremost as serious science writers. Film and radio provided new media where ideas and change in science could be explained in an entertaining way to the lay public.

Many popular science books debunk pseudoscience from the vantage point of a scientific understanding. One of the first – Martin Gardner's *Fads and Fallacies in the Name of Science* (Gardner 1957), first published in 1952 – also provides a flavour of the immense optimism surrounding science in the 1950s. Gardner's targets are interesting, representing a snapshot of then current conspiracy theories and applied scientific nostrums: flying saucers, dianetics and creationism are still with us; orgone theory, Lysenkoism and Babson's anti-gravity material have largely passed out of the public gaze. However, Gardner's assaults on osteopathy (a 'medical cult'), chiropractice (another 'medical cult') and the Bates method for correcting eyesight would raise eyebrows amongst medical practitioners today, many of whom recognize considerable medical value in such approaches. Many of the scientists writing about pseudoscience and the abuse of science are also somewhat hostile towards social sciences in general. Clearly many authors do not consider it worthwhile exploring the history of their own genre, and this explains, in part, why the same book has been written so many times by so many eminent people. Gardner's list of pseudoscience is echoed by Carl Sagan's in *The Demon-Haunted World: Science as a Candle in the Dark* (1996), which is, in turn, echoed by Robert Park's *Voodoo Science: The Road from Foolishness to Fraud* (2000) and *Superstition: Belief in the Age of Science* (2010), Michael Friedlander's *At the Fringes of Science* (1998), Michael Shermer's *The Borderlands of Science: Where Sense Meets Nonsense* (2001) and *The Believing Brain: From Ghosts and Gods to Politics and Conspiracies – How We Construct Beliefs and Reinforce Them as Truths* (2012), and Massimo Pigliucci's *Nonsense on Stilts: How to Tell Science from Bunk* (2010). The number of popular science texts solely dedicated to debunking pseudoscience suggests that they form their own subgenre, although as we will see many other popular science texts will pick up the theme of science versus pseudoscience.

Looking at the construction and articulation of the discourse of science in contemporary popular science texts (see box 6.1) shows us how scientists understand their own project, how they view their relationship to non-scientists, and what they consider to be 'dangerous' forms of 'spurious' knowledge abroad. Scientists and those associated with the project of science writing in this genre express an almost total faith in the project, method and goals of science, rarely express any concerns about the application of science, and consider the 'ignorance' of science that is expressed by 'non-scientific' media to be a major social ill. However, we also see a dramatic tension emerging from this form of writing: scientists writing for a popular audience will often generalize from their own specialist discipline to discuss the project of science as a whole. When they do this, they reinforce not only their own discipline's understanding of what science is, but also the fact that there are significant disjunctures between the different sub-disciplines of science.

Box 6.1 Popular science, Foucault and discourse

'Discourse' has come to take on a special role in contemporary social analysis. Rather than just being an aspect of speech that articulates ideas, discourses are seen as being *productive*. The idea of discourses making objects come into being stems from the work of French post-structuralist Michel Foucault. His ideas on discourse are presented in different ways in most of his works, so finding a single definition and way of considering discourse from a Foucauldian perspective is difficult, and this is made more complex by a proliferation of interpretations of his work that followed his death in 1984. However, his theory of discourse is highly relevant to considering the discourse of popular science in our society.

In 'Orders of Discourse' Foucault argues that '[I]n every society the production of discourse is at once controlled, elected, organised and redistributed according to a certain number of procedures whose role is to avert its powers and its dangers, to cope with chance events, to evade its ponderous, awesome materiality' (Foucault 1971: 8). That is, in order for discourses to achieve their effects they must be partially hidden from us such that we will not see their power to produce, or to position us as receivers of the discourse in quite specific ways. For Foucault, discourses are controlled by three categories of rules.

The first operates on the exterior of the discourse, excluding certain things such that the discourse has a specific effect. Desire is excluded, reason and folly are opposed, and truth and falsity are placed in opposition to each other. The discourse of science in our society, especially as articulated in popular science, clearly exhibits these features. Its effect, in this instance, is to construct the opposite of science as 'foolishness'.

> The second operates inside the discourse, classifying knowledge precisely (often according to disciplinary boundaries) and specifying who the legitimate authors of knowledge are. Again, the discourse of science that is articulated in popular science does precisely this, in many cases naming the authors who are legitimate (or at least classifying them – formal scientists) and presenting formal scientific disciplines as bounded and hierarchical.
>
> The third rule specifies who can employ the discourse and what conditions must pertain when the discourse is being used. Whilst some commentators using the popular science discourse have a clear 'right' to do this (Richard Dawkins, Brian Cox and others who have an accredited position in formal science), others will need to be admitted and allowed to use the discourse (lay commentators will often point to and highlight their ignorance of formal science when using the popular science discourse).
>
> Popular science clearly operates according to the rules Foucault identifies. The effect is complex. The discourse positions audiences outside science, constructs them as foolish (if they are not familiar with the knowledge being presented) and constructs science as single and unitary. The discourse also constructs *the* scientific community as external to society and comprising people with special capabilities. The discourse is flexible and capable of change; if we look at popular science across time we can see significant shifts in terms of the objects it names, the other it constructs and the effect it has on audiences.

In addition to their providing us with insights into how science looks at itself, we can also see popular science texts (I'm using a cultural studies definition of texts which includes all forms of cultural productions that we can 'read': books, TV science documentaries, popular science magazines, kids' TV programmes, etc.) as significant resources used in the social construction of science by the public. All of us are involved in constructing the meaning of 'science' and we do this with the resources at hand, a significant one of which is popular science accounts. We can learn a lot from looking at and analysing popular science.

Contemporary popular science: enforcing the 'ideology' of science

An iconic example of the popularity, and persistence, of popular science writing is Stephen Hawking's *A Brief History of Time* (Hawking 1988). This topped the non-fiction bestseller lists in both the USA and the UK for many weeks following its publication, selling more than

9 million copies; it was in the *Sunday Times* non-fiction bestseller list for over four years; and at the time of writing (March 2015) it has been translated into forty languages. Hawking's book, which is an account of recent breakthroughs in astrophysics and cosmology, is particularly significant in that it was written for a general audience by a practising and immensely successful scientist (Hawking is the former Lucasian Professor of Mathematics at Cambridge University, a post that his publishers and website point out was once held by Sir Isaac Newton, and is currently director of research at the Centre for Theoretical Cosmology at Cambridge). Hawking offers us a clear and unambiguous definition of his view of what the project of science is early on in *A Brief History of Time*: 'The eventual goal of science is to provide a single theory that describes the whole universe' (Hawking 1988: 10).

Hawking is a cosmologist: he produces theories of how the universe is constructed and what laws govern its behaviour. His understanding of what the project of science is is predicated upon his desire to reconcile the two partial theories that dominate physics, astrophysics and cosmology at the present time: quantum mechanics and the theory of relativity. Hawking's definition of the goal of science shows how theories that dominate the practice of science in a specific discipline will lead us to understand the project of science as a whole in a particular way; it also shows us how science makes sense of its own project. Hawking continues: 'Today scientists describe the universe in terms of two basic partial theories – the general theory of relativity and quantum mechanics' (Hawking 1988: 11).

As we saw in chapter 5, the conflation of all people working in formal science into a unified group of 'scientists' can be misleading, but is a move that the contemporary media make very frequently. Here, Hawking's use of the term 'scientists' is misleading in a similar way. It expresses a familiar – and not itself scientific – hierarchical analysis of the nature of science that is particularly common: science is great, but physics is the best (cf. Hacking 1983: 3). Hawking continues his account of what science is by subsuming all scientific disciplines into a unified project and set of perspectives:

> Now, if you believe that the universe is not arbitrary, but is governed by definite laws, you ultimately have to combine the partial theories into a complete unified theory that will describe everything in the universe. . . . Humanity's deepest desire for knowledge is justification for our continuing quest. And our goal is nothing less than a complete description of the universe we live in. (Hawking 1988: 12–13)

On this account scientific understanding as a whole ultimately rests on how physics makes sense of the world, and understanding all the laws of the physical world will provide the explanatory framework for all branches of science. This may, of course, be true. It is incontrovertible, however, that at this moment in time no such framework could possibly provide us with an appropriate way of understanding, for example, the workings of the human mind, the complex operation of social structure or the varied range of human social behaviour. All of these are aspects of the universe: whether we choose to see them as governed by definite laws is likely to be a product of, amongst other things, the gender, sexuality and cultural background of the observer. There is something deeply masculine about this account of the triumphal and progressive project of science.

In Hawking's account we see a restatement of the 'standard, formal' conception of the goal and aim of science. Science is an activity that will provide us with complete knowledge of the universe that we live in through constructing and testing theories. Science is comprised of a range of disciplines that are of varying degrees of 'hardness', with physics at the hardest and most rigorous end. Physics is the discipline that underpins all other scientific activities. It is also the discipline that made the greatest impact on the twentieth century through its association with creating atomic and nuclear weapons. It is the most male dominated and the most phallic of sciences. It is also, according to a more recent book co-written by Hawking, now the only valid form of knowledge in society and it is worth reminding ourselves of Hawking and Mlodinow's claim that the fundamental questions of being, meaning and reality were, formerly, questions for philosophy, 'but philosophy is dead. Philosophy has not kept up with modern developments in science, particularly physics. Scientists have become the bearers of the torch of discovery in our quest for knowledge' (Hawking and Mlodinow 2011: 13).

Hawking's *A Brief History of Time* was not the first popular science book that sought to bring science from the highest level of achievement into the life of the non-scientist reader. I've already mentioned Watson's book on DNA and Einstein on relativity; Werner Heisenberg – inventor of the uncertainty principle – produced a number of accounts of quantum mechanics for a popular audience (Heisenberg 1958, 1959), and Richard Feynman, Nobel laureate 1965, was, by the time of his death in 1988, a well-known and popular commentator on a range of social and scientific matters (Feynman 1999, 2000). Popular science, particularly that produced by famous scientists, is popular.

Unlike Einstein's, Heisenberg's and Hawking's, Feynman's popular science publications deliberately enter the territory of the world outside physics. He enters into debate on a range of social and cultural topics, but does so from the perspective of a scientist who is looking in on the social and cultural world to offer an 'external' critique. In a series of public lectures titled 'A Scientist Looks at Society', delivered in 1963 (Feynman 1999), he discussed how society was dependent upon science in many ways, but profoundly ignorant of science at the same time – a familiar theme today. In keeping with the style of this genre he provides us with an initial definition of science:

> What is science? The word is usually used to mean one of three things, or a mixture of them. I do not think we need to be precise – it is not always a good idea to be too precise. Science means, sometimes, a special method of finding things out. Sometimes it means the body of knowledge arising from the things found out. It may also mean the new things you can do when you have found something out, or the actual doing of new things. This last field is usually called technology. (Feynman 1999: 4–5)

A deceptively simple definition, and one that places technology as a cornerstone of the project of science. Feynman may be slightly mislead-ing here: he is, after all, giving us a popular definition of what science is for use in a popular science context. Unlike other authors, he is writing about how science can be applied to aspects of the social world not normally considered 'scientific'. Thus his definition of science has to be open textured, rather than the tightly and rigidly defined versions of science that, say, physicists present to their lay audiences when explaining physics. B. K. Ridley's book *On Science* (Ridley 2001) is a general overview of the scope and limits of science in contemporary society, written from the perspective of a physicist (Ridley is professor of physics at the University of Essex). Ridley presents an in-depth and critical account of scientific thinking and, unlike some other contem-porary commentators on science, avoids the temptation to present science as the best form of thought available, and the most suitable solution to all problems. However, like Hawking, he sees science as essentially reducible to the paradigm of physics: 'Science is essentially a description of the motion of matter' (Ridley 2001: 31). This description of the 'essence' of science is certainly debatable, and for our purposes as social investigators of science we would want to challenge it in a number of ways. However, Ridley is expressing his own thought com-munity's understanding of what science is: the science that physicists do is, ultimately, a science of describing the motion of matter (exactly the same point that Thomas Hobbes made in *Leviathan* ([1651] 1968) – see chapter 4). In Ridley's case, his thought community of physicists reduces phenomena to fundamental units and forces, and his definition of science (Ridley is talking about science *as a whole*, not simply the science that he and his colleagues carry out) thus becomes an atomistic, reductionist one.

The reductionist view of science is the current dominant model for the project of science as a whole (Lewontin 1993; Midgley 2010). Scientific explanations look for ultimate causes at the level of fundamental and discrete units: the cause of cancer is our genes, the structure of matter is ultimately caused by fundamental particles. The correctness or otherwise of such perspectives is not at issue here, although, particularly in the case of the dominant genetic reductionism in contemporary biological science, there are some significant dissenting voices. However, such reductionism does, quite neatly, match up to the understanding of the social world as being the manifestation of the properties of individual human beings (Lewontin 1993: 107ff). Inside the genre of popular science we can see a similar form of reductionism in terms of the topics and titles produced by authors. Of course, there are still some general texts that analyse science as a whole, as we have noted, but many of these do this from a specific disciplinary perspective. However, since the early 2000s the expansion of 'atomized' popular science texts has been quite remarkable: the single-word book title is very popular in popular science. Just looking across the shelves of the 'popular science' section in my local (medium-sized) bookshop produced the following list from authors between A and H: *Genius, Elements, Genetics, Nothing, Chaos, Brain, Smarter, Anatomies, Quantum, Higgs, Branches, Curiosity, Unnatural, Shapes, Flow, Digitized, Perv, Spectrums, e = mc², Gravity, Neutrino, Antimatter, Paleofantasy, Abundance, Dirt, Extremes, Survivors, Life, Discord, Volcano* (and I've excluded three one-word-titled biographies of scientists). Chopping the world up into discrete units for consumption is a strategy that appeals to popular science publishers.

But is Ridley's science the same science as the palaeontology (the discipline that deals with the life of previous geological periods as found in fossil remains) that palaeontologists are doing? Palaeontologist Stephen Jay Gould, author of a significant number of popular science books, as well as a distinguished naturalist in his own right, expresses concerns about some forms of popular science writing.

> I have fiercely maintained one personal rule in all my so-called 'popular' writing. (The word is admirable in its literal sense, but has been debased to mean simplified or adulterated for easy listening without effort in return.) I believe . . . that we can still have a genre of scientific books suitable for and accessible alike to professionals and interested laypeople. The concepts of science, in all their richness and ambiguity, can be presented without any compromise, without any simplification counting as distortion, in language accessible to all intelligent people. (Gould 2000: 16)

Gould's writings on science tend to focus on his own disciplinary specialism: his concern is to defend his perspective on the concepts and main tendencies of his discipline and to promote the way they are understood by participants in that project, and by the wider general public. He, too, is talking about science as a whole, but only in the sense that palaeontology and biology are a part of that wider project. Gould is a subject-specific scientist writing about his own small area of science: his presentation of science begins from his own position inside the research project of science as a whole, hence the emphasis on the explication of evolution as a concept, rather than an emphasis on the nature of, for example, empirical work in general. Despite this there are consequences emanating from his writing about his discipline for our understanding of science as a whole.

This is in marked contrast to a group of writers who are perhaps best described as 'fans'; journalists and others who write on science from outside an esoteric thought community. Once again, we find a strong thread of opposition to 'anti-science' (which encompasses religion, social science and imputed ignorance) in these texts, and a very strong validation of science as the single best form of knowledge available. Bill Bryson, an author best known for his travel writing, wrote a bestselling 'rough guide' to science in 2003. From meetings and interviews with scientists – formal and informal – Bryson constructs a picture of science that is full of information, but also captures his and others' wonderment at the scientific endeavour. He states his mission as being 'to see if it isn't possible to understand and appreciate – marvel at, enjoy even – the wonder and accomplishments of science at a level that isn't too technical or demanding, but isn't entirely superficial either' (Bryson [2003] 2004: 24). Bryson does present a really engaging account of science and one which, unsurprisingly, is entirely uncritical of any aspect of the scientific endeavour. And why should he be critical; like most science writers in this genre, Bryson will not have come across any kind of critical analysis of science or scientism. No wonder he produces a 'gee-whizz' account of science: after all, it really is just amazing what science has achieved.

In similar vein, and addressing the same audience, comedian (and former physicist) Ben Miller's *It's not rocket science* (2012) presents a light-hearted and slightly irreverent picture of a few of the most high-profile (or as Miller puts it 'exciting') issues in contemporary science, including climate change, cosmology and genetics. The book repeatedly tells the reader not to worry about learning things, or trying to understand very difficult concepts; the author reassuringly states that the whole point of the book is to provide fun and to show how exciting science is. Indeed, the book's cover states 'Discover the surprisingly simple ideas behind the most exciting bits of science'. Miller continually expresses his deep love for science, his admiration for scientists and their wonderful methods, and the superiority of science, and particularly physics, to other modes of explanation. In an almost obligatory passage for this type of book Miller presents the publicly aired fears about the Large Hadron Collider (LHC) (there were stories in the media that the collider could present a danger to humans through generating black holes) as pseudoscience and places this in direct contrast to science.

> The scientists at CERN are far too polite to say it, but if you ask me, what the LHC-black-hole controversy boils down to is the difference between science and pseudoscience. One of the main problems we have in our culture is that the media, in general, can't tell the difference. In some ways that's understandable; after all, pseudoscience, by definition, makes its way in the world by mimicking science and sometimes manages to do quite a good job. And when you take into account the fact that most people who work in the media have arts degrees and are positively phobic about science, you can see there's a weakness to be exploited. (Miller 2012: 42)

Who is to blame for the public's worries and concerns about CERN's LHC? The answer is that it is the public themselves (with their 'febrile imaginations' (ibid.: 43)) and the arts-graduate-dominated media who are to blame, and certainly not the scientists who run the machine. Yet it could be argued that the CERN scientists do have a duty to explain what it is they are doing to the public who (1) have funded the project and (2) are always being told that they need to know more about science. Miller, in a footnote, notes that the LHC team did actually write a 100-page 'watertight refutation of the proposition that a microscopic black hole might destroy the planet. . . . [B]ut reading it can't help but leave me, for one, with the nagging question: Is this the best way for the most talented particle physicists in the world to be spending their time?' (Miller 2012: 42 n. 6). That's quite astonishing: according to Miller the public, who provided the £2.6bn to build the LHC (Science & Technology Facilities Council, http://www.stfc.ac.uk/646.aspx), should not expect the physicists at CERN to waste their time trying to explain the thing or reassure them about their concerns. But, presumably, that leaves the public at the mercy of the 'arts-graduate-dominated' media for their information?

Bryson and Miller do similar things in their books, and it is welcome to see science being presented in a lively and, particularly in Miller's book, humorous way. However, once again we are presented with a version of 'science is great and physics is the best', a common theme in contemporary science journalism. We can speculate as to why this may be the case, and an obvious explanation comes from even a cursory reading of journals such as *New Scientist*, *Scientific American*, *Nature* or *Science*. Here we see science being presented to a knowledgeable audience, although readership, particularly for the first two, extends beyond the boundaries of those working in formal science. Science does not need to be justified or explained for this audience, but this audience does not want to see science being deprecated or even strongly questioned: these journals are 'doing a job' of defending science from irrationality and prejudice, and fulfilling a 'necessary' function of presenting science in the strongest possible light to ensure the continued adherence of states, corporations, educators and the general public to the project of science.

Undergraduate students on my science, culture and society course at the University of Brighton analysed formal science journals such as *Nature* and *Science* and concluded that these were not deliberately patronizing or excluding, but did assume a vast stock of prior knowledge which can produce an effect in readers of feeling patronized or excluded from the apparently most important aspect of contemporary industrial societies. Such results are more likely to be the product of the needs for economical use of language leading to the journalists making quite large assumptions about the knowledge base and capabilities of the readers. The result is a reinforcement of the 'separateness' and 'difficulty' of science for those not familiar with the discourse being presented. This is likely to be compounded with the almost relentless assault on the arts, humanities and social science one finds in these texts. We've already seen what Sir Paul Nurse thinks of social science, and Ben Miller is equally scathing.

Ben Goldacre's 2008 bestseller *Bad Science* presents a very useful guide to how drug companies, in particular, will distort and selectively choose data to support particular claims they are making. Goldacre also debunks claims made by TV nutritionist 'Dr' Gillian McKeith, vaccination scaremongers and homeopaths. The book is a solid attack on pseudoscience and on distortions of the standard scientific method, and it provides readers with tools to make more informed assessments of media and other reports. It is, perhaps, rather prone to scaremongering in that it suggests that science is being ignored. 'Today, scientists and doctors find themselves outnumbered and outgunned by vast armies of individuals who feel entitled to pass judgment on matters of evidence – an admirable aspiration – without troubling themselves to obtain a basic understanding of the issues' (Goldacre 2008: x). Really? A 'vast army'? As with Nurse's argument given earlier, I think it is fair to say that this is not a scientific proposition, and not even a sensible proposition. And throughout the book Goldacre, like Miller, pours scorn on humanities graduates. For example:

> After all, as any trendy MMR-dodging north-London middle-class humanities-graduate couple with children would agree . . . (ibid.: 277)

and:

> the people who run the media are humanities graduates [the same ones identified by Miller, presumably] with little understanding of science, who wear their ignorance as a badge of honour. Secretly, deep down, perhaps they resent the fact that they have denied themselves access to the most significant developments in the history of Western thought from the past two hundred years; but there is an attack implicit in all media coverage of science: in their choice of stories, and the way they cover them, the media create a parody of science. (ibid.: 207–8)

and:

> humanities graduates in the media, perhaps feeling intellectually offended by how hard they find science. (ibid.: 271)

Again, these are not scientific propositions being put forward by Goldacre, are based on no media analysis or firm evidence whatsoever and are barely even sensible statements. If a humanities graduate were to make statements concocted in a similar way about, say, an enzyme or a vitamin Goldacre would take them to task. Insulting the audience you most want to convert to your position seems perverse, but is an inevitable consequence of articulating the strong scientism we see throughout popular science discourse. Scientism rests on setting scientific knowledge above other forms of knowledge; it has to sustain this.

Popular science's regular attacks on the arts, humanities and social sciences are predictable and, in a world where scientism dominates formal thought, inevitable. It seems clear that the attack on the credulous is a trope that has to be deployed in this genre. However, as popular science as a genre has expanded in recent years an additional target has been selected: religion. The most obvious example of this is Richard Dawkins's *The God Delusion* (2006), which had sold over 3 million copies by late 2014. Dawkins is a vociferous campaigner against religion in all forms, and particularly against religion being taught in schools or having any kind of influence or interference in science. Dawkins would be overjoyed to see religion banished or banned; he is quite convinced that religion is pretty much the root of all evil in the world:

> Imagine ... a world with no religion. Imagine no suicide bombers, no 9/11, no Crusades, no witch-hunts, no Gunpowder plot, no Indian partition, no Israeli/Palestinian wars, no Serb/Croat/Muslim massacres, no persecution of Jews as 'Christ-killers', no Northern Ireland 'troubles', no 'honour killings', no shiny-suited bouffant-haired televangelists fleecing gullible people of their money ('God wants you to give until it hurts'). Imagine no Taliban to blow up ancient statues, no public beheadings for blasphemers, no flogging of female skin for the crime of showing an inch of it. (2006: 23–4)

This is a rather strange theory of history; not really a theory as such, but a kind of scientistic reductionism, a positivist view of history where the only causal factor in war and conflict is religion. Significantly, here, Dawkins has not stopped to consult historians about the plausibility of this thesis (the two largest conflicts of the twentieth century – the world wars – were not 'religious' conflicts, but World War II was prompted by a clash of 'scientific' ideologies), but then, like other popular science texts, *The God Delusion* has a very dim view of the arts

and social sciences. Despite religion being a wholly social phenomenon, Dawkins does not examine any sociology of religion, preferring
to construct his own theories of why religion persists in society. The
reason is that people of faith are simply foolish or ignorant of the truth
of science: '[F]or many people the main reason they cling to religion
is not that it is consoling, but that they have been let down by our
educational system and don't realize that non-belief is even an option'
(ibid.: 22).

Prior to *The God Delusion*, Dawkins was best known for his 'selfish
gene' theory, a consistently reductionist account of life which sees
organisms as merely carriers for DNA molecules that are determined to
propagate themselves at any cost (Dawkins 1989). Selfish gene theory
has been very successful, but as Dorothy Nelkin points out it has been
a factor in producing a form of language in the biological sciences that
we can call 'God talk'. 'In popular writings, they are using spiritual
constructs and religious rhetoric to describe their work and to convey
its significance' (Nelkin 2004: 139). God talk persists because:

(1) It is a reflection of an ethos that has long driven scientific pursuits.
(2) It also reflects an enthusiasm about science that is often expressed in
    excessive hyperbole.
(3) It reflects the wider prevalence of God talk that pervades political
    rhetoric in the United States.
(4) It appears to be an instrumental response to public concerns about
    the social and ethical implications of contemporary biology. (Nelkin
    2004: 140–1)

The outcome of this is that biology is treading on the terrain formerly
considered to be the province of religion. 'DNA has taken on the social
and cultural meaning of the soul' (Nelkin 2004: 145). The reductionist
approach that sees everything as explainable by genes and DNA has
led to the position where DNA is seen as being immortal (like the soul),
and has led to bitter conflicts between religion and biology. The gene
is a powerful deterministic agent: this meets the needs of people for
having a grounded understanding of what contemporary life is: I am
my genes. We define ourselves through the genome – it is, in Nelkin's
words, a solid and immutable structure.

Although Dawkins expresses a strong, science-based hostility
towards religion, others in the scientific community are more concilia-
tory and there has been, throughout modernity, a kind of truce between
science and religion, where each sticks to its own area of interest:

> Science is about causes, religion about meaning. Science deals with how
> things happen in nature, religion with why there is anything rather
> than nothing. Science addresses specific questions about the workings
> of nature, religion addresses the ultimate ground of nature. (American

Association for the Advancement of Science 1995, cited in Nelkin 2004: 144)

and:

> Questions of meaning and religion emerge from our deepening under-standing of the natural order. Issues of value and meaning are grounded in the disciplines of ethics and religion. The scientific community needs to be in dialogue with both fields in order to understand the cultural context within which science operates and to respond to the societal issues opened up by scientific discovery and technological development. AAAS provides a uniquely credible forum for that engagement because of its disciplinary breadth. (American Association for the Advancement of Science, 13 July 2014, http://www.aaas.org/page/doser-overview)

This demarcation of areas of interest and mutual recognition of value in the other has come under significant pressure, and primarily in the area of biology. The reductionist doctrine, of life to DNA and evolution, has led to a backlash from conservative religious fundamentalists, and this has been particularly visible in the USA since the mid-1990s. The argument has focused on the very important topic of whose theories will be used in schools to teach children about the origins of life: creationism or evolutionary theory. Creationism is, in its pure form, the belief that the Christian Bible's representation of the creation of life on Earth is a true and accurate account: any other account (such as, say, Darwin's theory of evolution) is a crude lie that not only presents a falsehood but also is a sacrilegious position that insults Christianity and maligns the role of faith. This is quite an extreme position and a difficult one to hold in the face of almost insurmountable evidence against it. Arguments between Darwinists and creationists peaked in the early part of the twentieth century, and until the 1990s evolutionary theory was the prime mode of explanation for life on Earth in biology classes across the USA and UK. In more recent years a number of attempts have been made to reconcile creationism with contemporary scientific evidence and practices. The most successful of these is intelligent design theory (ID or IDT), which has been rather rudely called 'creationism with a website'. IDT is the project of establishing 'by the usual scientific appeals to reason and evidence' that the world and life in it were purposefully designed by an intelligent agency competent to the task of creating a universe. ID looks for points of 'irreducible complexity' in biological systems and uses these as evidence that a standard evolutionary account cannot explain their emergence: complexity implies deliberate design. Leading ID theorist Michael Behe uses the example of a mousetrap as an irreducibly complex system which could not have arisen without the intervening hand of a designer, and draws an analogy with the internal structures of the mammalian cell to suggest that it, too, must have been designed

(Behe 2003: 294, 301). Needless to say, molecular biologists disagree strongly and are scathing about IDT 'experiments'.

Perhaps surprisingly IDT has been lent support by a leading STS scholar, Professor Steve Fuller of Warwick University (Fuller 2007). Fuller describes himself as a secular humanist who is not trying to promote, or denigrate for that matter, any religious cause or position. His main thesis is that IDT should be allowed the same position in science as any other theory and should be allowed to work towards proving that some kind of intelligent designer was responsible for the generation of life on Earth. This should be visible in experiments and observations. For Fuller, IDT and evolutionism should be discussed at the same level (i.e. the general), as evolution, at the general level, has no more explanatory power than IDT. Perhaps this is the case – evolution can't be 'proved' in the way that the theory of gravity can be. But IDT can't be proved either. For Fuller this makes them equivalent.

However, when you look at IDT science writing – reports of experiments and formal observations – they are quite strange objects, almost a simulacrum of what other scientists get up to. Interestingly, almost all IDT studies are funded by US-based fundamentalist Christian churches, which does suggest some wider public engagement with science, although perhaps not the kind of public engagement that many STS commentators have called for.

Fuller's point, that IDT and Darwin's theory can be treated in the same way, is debatable, but his underlying concern about science in society deserves further analysis. As far as he is concerned, a scientific paradigm that places the randomness of evolution at its centre serves a social purpose of reducing the value of human life to that equal to all other living things; this paradigm is a necessary consequence of new practices in the life sciences, particularly cloning, genetic testing, stem cell research and genetic modification of organisms. There's nothing special about us from the perspective of evolutionary theory, according to Fuller, a position shared with many religionists (*and* humanists) who find the mechanistic reductionism of selfish gene theory to be abhorrent. Fuller also notes that science emerged from religion and to see them as opposites rather than as complementary is wrong. This point is well established in sociology of science:

> We note that religion directly exalted science, that religion was a dominant social force, that science was held in much higher social esteem during the latter part of the [seventeenth] century, and on the basis of much corroborative evidence we conclude that religion played an important part in this development. (Merton 1973: 249–50)

Can there be a resolution of the hostilities between popular science and religion? On the present evidence it seems unlikely:

popular science texts are intolerant of religion, most people of faith in UK society seem remarkably unconcerned about this, and the very small number of religious fundamentalists in UK society do not want any rapprochement with science, which they simply consider to be falsehoods anyway. But we could imagine some kind of truce re-emerging if popular science were to study religion in a more sensible way. Rather than seeing it as codified foolishness, sociology of religion shows that faith is largely superfluous when understanding religion (Peter Berger's classic 1973 book describes how sociologists, from Marxists to functionalists, understand this); what we need to see is the tradition, community and values that religiosity provides:

> If you're an atheist, I can heartily recommend involvement in religion. It offers a sense of belonging and it offers tradition, which can be reassuring and comforting. It offers discipline, teaching us that there is something outside ourselves to which we should bend our personal will. If we do it right, religion helps us lead better lives, with a commitment to justice and social action. Sociological research shows that involvement in organised religion is good for our health and well being. (Tom Shakespeare, *A Point of View*, BBC Radio 4 broadcast, 24 May 2014, http://www.bbc.co.uk/news/magazine-27554640)

A more systematic and informed approach to understanding religion might promote a slightly more tolerant attitude and a less strident reductionism. And this might begin to address the concerns that many in the lay public have about science in society: that it is beyond control and not necessarily working in the best interests of society as a whole. This suspicion of science is understandable when leaders of the scientific community such as Paul Nurse and Richard Dawkins show such disregard for the attitudes and opinions of the general public. The assumption that science is just great, and all it needs is to explain itself better, is a dangerous one. It assumes that there is only one valid form of knowledge, and that this knowledge can somehow be wholly objective. As Evelyn Fox Keller pointed out in an article in *New Scientist* on the controversy surrounding 'Climategate' (where hackers, in November 2009, broke into a leading UK climate research centre and found emails they claimed showed scientists manipulating and covering up data), 'If they [climate scientists] are to be blamed at all it is for adhering to an image of science as capable of delivering absolute (and value-free) truth: an image most scholars recognise as indefensible, and one that, among themselves, most researchers accept as unrealistic' (Keller 2011: 22). But it isn't just the lay public who are remiss here. Formal science communities can also be accused of being foolish and credulous when we look at the troubling issue of scientific fraud and misconduct.

Scientific fraud

The standard account of science, with its emphasis on objectivity, selflessness and disinterested appraisal of results, suggests that scientific misconduct and fraud will simply not take place: scientific communities are self-regulating entities that will ensure that only the truth emerges. Yet it remains the case that scientific misconduct and even fraud take place across the formal scientific community, although estimating the extent of this is difficult.

Scientific fraud strikes at the heart of the ethos of science: Merton's institutional imperatives (see chapter 5) imply that fraud should simply not happen; Weber's idea of science as a vocation ([1918] 1989) also implies that the search for the truth is what is inspiring scientists, not the quest for fame or material reward. But the fact that governments and scientific institutions feel a need to define what constitutes fraud and misconduct suggests that this is an endemic problem. The UK government Department for Innovation, Universities and Skills issued a two-page ethical code to scientists in 2007 titled 'Rigour, respect, responsibility: a universal ethical code for scientists'. The aim of the code was:

- to foster ethical research
- to encourage active reflection among scientists on the wider implications and impacts of their work
- to support constructive communication between scientists and the public on complex and challenging issues.

In the United States the National Institutes of Health identify research misconduct as 'fabrication, falsification and plagiarism', and many researchers would identify these three types of activity (making things up, manipulating data or accounts of process, stealing others' work) as the main forms of academic fraud. We should add another one: selectively picking data or deliberately excluding data that is inconvenient. How prevalent are these activities? It is difficult to determine, but Daniele Fanelli's meta-survey of studies into research misconduct found that:

> on average, about 2% of scientists admitted to have fabricated, falsified or modified data or results at least once – a serious form of misconduct [by] any standard – and up to one third admitted a variety of other questionable research practices including 'dropping data points based on a gut feeling', and 'changing the design, methodology or results of a study in response to pressures from a funding source'. In surveys asking about the behaviour of colleagues, fabrication, falsification and modification had been observed, on average, by over 14% of respondents, and other questionable practices by up to 72%. (Fanelli 2009: 8)

Ana et al. followed up this study to look at research in low- and medium-income countries; they found largely similar results (Ana et al. 2013). Worryingly, they found little or no discussion of research misconduct in the countries they included in their study, and most developing countries, with the exception of China, had no systems for responding to scientific misconduct: 'There is an understandable tendency to deny research misconduct, and most countries seem to take decades of debate and exposure of cases of misconduct to accept the need for a systematic response' (ibid.: 5). In terms of reporting scientific misconduct, again it is difficult to estimate how much is taking place. However, *Nature* estimated in 2013 that retractions of scientific papers, where a journal is informed of, or realizes itself, that a paper it has published contains errors or deliberate falsification of results, had 'increased tenfold during the past decade, with many research studies crumbling in cases of high-profile research misconduct that ranges from plagiarism to image manipulation to outright data fabrication' (Yong et al. 2013: 454). This rise in the number of retractions may be partly due to the continual growth in the number of scientific publications (see box 6.2).

Box 6.2 The growth of scientific knowledge productions

The amount of scientific knowledge produced has expanded dramatically in recent decades; this can be seen in the number of journal articles published. Predictions made in the early 1960s that the number of scientific papers published in journals would double every 10 to 15 years have proved correct, although there is variation between disciplines. For example, chemistry doubles over a 20-year period, maths over a 22-year period, but technology-oriented papers double in a 9-year period (Larsen and von Ins 2010). Older, well-established disciplines such as mathematics and physics tend to have slower growth rates than new disciplines, including computer science and engineering, but the overall growth rate for science has been at least 4.7 per cent per year, which gives a doubling time of 15 years (ibid.: 600).

There is, at least, growing discussion about the problem in scientific publications, and journals are now quick to retract studies where problems or inconsistencies are identified. But the question that is really pertinent is why these studies are published in the first place. Scientific papers are subject to considerable pre-publication review and scrutiny to ensure that they are correct and accurate (see box 2.3 on p. 36). After all, a journal needs to protect its reputation, as well as having a duty to the wider public to publish only accurate and honest research. So how does bogus material get through?

One reason is that all formal scientists are now under tremendous pressure to 'publish or perish', and to publish in the highest-ranked journals. This means that the original work leading to a publication may not be checked as thoroughly as it should be. It also may be the case that formal scientists feel under pressure to, perhaps, be economical with the truth in terms of presenting data, or hiding negative results. Alongside this we need to recognize that peer reviewing journal articles, whilst a very important activity for the integrity of formal science publications as a whole, may not be the top priority for all formal scientists given the terrific pressure to produce and publish their own work. In research I carried out some time ago, I interviewed the head of a laboratory who told me that he often got his PhD students to do peer reviews for him as it was good experience for them, and he was too busy. We also need to remember, as we have seen from the example of molecular biology work in chapter 2, that much formal science laboratory work is extremely specialized, and even inside an esoteric thought community a peer reviewer may not be a complete expert on the work they are reviewing.

A second reason is that science is agonistic, and even inside esoteric thought communities there will be competition and secrecy. A team of researchers will want to be the first with a new discovery and will do their utmost to avoid being beaten to publication. This can, occasionally, lead to journals presenting inaccurate or fabricated material. For example, a paper published in *Biochemical and Biophysical Research Communications* in July 2013, reporting that over-expression of two novel proteins in fat cells leads to improvements in metabolic processes related to diabetes and obesity in mice, was co-authored by five members of staff at the University of Thessaly in Greece. The problem here is not with the research findings – most probably correct – but that these members of staff are unknown at their university and no trace of them as researchers anywhere can be found. The paper was plagiarized from another cell biologist, Bruce Spiegelman at Harvard Medical School, and submitted to the journal as a spoiler against a rival laboratory's work (Butler 2013). It has since been withdrawn permanently.

Speed can also contribute to errors entering honest scientific research. A blockbuster paper reporting the creation of human embryonic stem cell lines from cloned human skin cells was published in 2013 in the prestigious journal *Cell*. It was accepted for publication just three days after receipt, and was published online twelve days later (Cyranoski 2013). This is a very quick turnaround time, but for work in such a competitive field as human stem cells it is likely that authors and publishers will do their best to get work into print to avoid being beaten by rivals. Unfortunately, some of the images used in the paper were duplicates and, in addition, had misleading labels. The lead author, eminent reproductive biologist Shoukhrat Mitalipov, dismissed the errors as 'typos', and no one is accusing the team of anything other than sloppiness. However, the human stem cell research community have form on exactly this topic and with exactly these errors. 'The last time the same feat was claimed – by then Seoul University professor Woo Suk Hwang – duplicate images were noted anonymously and the breakthrough was later debunked' (Cyranoski 2013: 543) (see box 6.3).

Box 6.3 Woo Suk Hwang

Woo Suk Hwang suffered a spectacular fall from grace in 2006 when he was dismissed from his post as professor of biotechnology at Seoul National University. Hwang came to prominence through work on human stem cells and cloning; he is best known for cloning the first dog – Snuppy – in 2005, but his most significant work, showing that cloned human cells can become 'pluripotent' stem cells, was fabricated, and was retracted by the prestigious journals which had published it. This brought other aspects of Hwang's work into public scrutiny; he was accused, initially, of forcing a graduate student in his laboratory to donate human eggs for experimentation. Subsequently, in 2006 Hwang and three other members of his team were charged with fraud, embezzlement of some $3m for Hwang's own personal use and for the illegal purchase of human eggs for use in experiments, and breach of Korea's bioethics laws. Hwang was finally sentenced in 2014 to eighteen months' imprisonment, but to serve time in jail only if he breached the terms of his probation. After this he returned to work, offering a bespoke service to clone your pet dog. His latest venture, at the time of writing, is a partnership with Shoukrat Mitalipov to break into the Chinese stem cell and cloning market (Cyranoski and Deng 2015).

Coincidentally, during the course of writing this chapter, yet another case of academic misconduct concerning human stem cells and the fabrication/duplication of a number of images came to light. *Nature* retracted two papers published in January 2013 reporting remarkable results: transforming adult cells into pluripotent stem cells (*Nature*, 511, 7507, p. 112). The retraction followed an investigation by the papers' authors' home institution which found one author guilty of misconduct, and found inadequacies in data management, record keeping and oversight. The journal showed admirable speed in retracting the papers, but admitted to its own shortcomings in allowing the paper to be published in the first place:

> Our policies have always discouraged inappropriate manipulation [of images]. However, our approach to policing it was never to do more than check a small proportion of accepted papers. We are now reviewing our practices to increase such checking greatly. . . . Underlying these issues, often, is sloppiness, whether in the handling of data, in their analysis, or in the keeping of laboratory notes. As a result the conclusions of such papers can appear misleadingly robust. . . . We need to put quality assurance and laboratory professionalism ever higher on our agendas, to ensure the money entrusted to us by governments is not squandered, and that citizens' trust in science is not betrayed. (Editorial, 'STAP Retracted', *Nature*, 511, 7505, pp. 5–6 at p. 6)

The point here isn't whether or not Hwang and his ilk fabricate their results. It is that formal scientists themselves – journal editors, peer reviewers, colleagues – could be fooled by specious scientific work. Lack of time, expertise, concerns for confidentiality are all factors here, but that doesn't excuse fraudulent work in the first place. The fact that fraudulent or error-strewn work is published in esteemed journals is of concern here: the problem is well known, and scientific journals, which in themselves are well funded, should know better. Bear in mind that these journals are presenting what they consider to be the truth, and would themselves, in all likelihood, be just as scathing about those who are ignorant of science as Sir Paul Nurse, Ben Goldacre and Richard Dawkins. But surely their publication of specious and fallacious material indicates at least a certain amount of credulity?

Social studies of science – despite being derided by eminent formal scientists – can suggest some lines of inquiry that might help us understand how it is that scientific fraud and error take place. To start with, social analysis of science suggests that formal science has fallen victim to its own 'ideology' of disinterestedness and universality; as a whole, formal science doesn't consider that scientific fraud and misconduct are that significant, and thinks that the institutional imperatives that it thinks govern science are self-correcting mechanisms that will prevent fraud and misconduct. Secondly, using a sociology of science and sociology of work approach we can see that formal science teams in the discovering sciences rarely, if ever, carry out experiments that replicate the work of others. As John Ionnadis notes, in a paper analysing the validity of medical research and titled 'Why Most Published Research Findings Are False':

> Several methodologists have pointed out that the high rate of nonreplication (lack of confirmation) of research discoveries is a consequence of the convenient, yet ill-founded strategy of claiming conclusive research findings solely on the basis of a single study assessed by formal statistical significance, typically for a p-value less than 0.05. Research is not most appropriately represented and summarized by p-values, but,

unfortunately, there is a widespread notion that medical research articles should be interpreted based only on p-values. (Ionnadis 2005: 696)

Formal scientists have other things to do than reproduce their colleagues' work: producing their own research and results so that they achieve the aims and milestones set by their funding body, and generating new research proposals for future funding being the two most important. On top of that, team members will be involved in the day-to-day administration of the laboratory, supervising graduate students and carrying out teaching duties. There is not much space left over. Yet ensuring that results are reproducible should be the corner-stone of ensuring that published results can be trusted. A 2013 op-ed piece in *Nature* noted that 'In recent years, it has become clear that biomedical science is plagued by findings that cannot be reproduced' (Russell 2013). Similarly, unless we know the social conditions that pertain inside laboratories, the ways that hierarchies are determined and enforced, the power relationships that exist inside esoteric thought communities, we will not understand why it is that academic miscon-duct perpetrated by individuals can emerge in what is, in most cases, a collectively produced product.

What do popular accounts of science show us?

Firstly, simply by their number and proliferation, they show us that science is a hugely popular topic in our contemporary culture. The sheer numbers of such books being sold shows a great desire on the part of the general public to know more about science. Secondly, popular science accounts serve to restate the triumphal status of science in contemporary society. These accounts rarely admit that science cannot explain everything. Frequently, where such failings are alluded to, as in Stephen Hawking's refusal to answer questions about God, science is seen as being the method by which such answers will be available in the future. Thirdly, science is presented as being a unitary phenomenon: many branches of science will be shown and discussed, but all are ultimately connected by common purpose and method, according to these accounts. Of course, individuals writing inside their own disciplines will often express their opinion that, say, physics or biology is the 'supreme' science, but, ultimately, science is represented as being a unified whole – a restatement of the essentialist picture of science. Fourthly, science is presented as the rational opposite of organized foolishness and credulity, the popular science definition of religion.

Finally, these accounts allow us to see the prevalence of a scientific worldview, an 'ideology' – scientism – that is a commonly shared perspective on the world around us (Midgley 2010). The public are presented with and assimilate a range of resources in constructing their understanding of science and, given the composition of much of this material, it is unsurprising that we see a societal understanding of science that reflects aspects of popular science accounts. Scientism suggests that explanations for the world around us should be based on scientific principles rather than religion, superstition or sheer guess-work. Writing in 2002 in celebration of Stephen Hawking's sixtieth birthday, and in praise of Hawking's refusal to answer 'god questions', popular science writer Michael Shermer couldn't avoid dropping into the register of 'God talk' (Nelkin 2004). Not only did he title the article 'The Shamans of Science', he also made some bold claims for Hawking's status:

> What is it about Hawking that draws us to him as a scientific saint? He is, I believe, the embodiment of a larger social phenomenon known as scientism. Scientism is a scientific worldview that encompasses natural explanations for all phenomena, eschews supernatural and paranormal speculations, and embraces empiricism and reason as the twin pillars of a philosophy of life appropriate for an Age of Science. (Shermer, *Scientific American*, June 2002, p. 25)

It is this scientistic worldview that is both cause and consequence of the status of science in Western industrial societies. There is little dissent amongst the scientific community, although professor of physics at MIT Ian Hutchinson's book *Monopolizing Knowledge* (2011) is a trenchant critique of the idea that science can be the source of all knowledge. For Hutchinson, scientism is 'busily, but largely surreptitiously, at work throughout practically the entire intellectual and cultural landscape' (preface, http://monopolizingknowledge.net/contents.html). It is perhaps unfortunate that this attack on scientism comes from a formal scientist who also positions himself strongly inside another thought community:

> In my intellectual journey as a follower of Jesus Christ and as a professional scientist, I came to believe a long time ago that the much-discussed tensions between science and Christianity are part of a wider disagreement between the improper extrapolation of science and – well – everything else. The improper extrapolation of science is approximately the belief that science, modelled on the natural sciences, is the only source of real knowledge. I call that belief scientism. (ibid.)

Popular science serves to reinforce scientism, and scientism provides the metanarrative to popular science. Yet there is complexity inside this: whilst the popularity of popular science texts suggests the public want to be informed, they also want to be entertained, and science is entertaining – particularly in its guise of science fiction, considered in the next chapter.

Further reading

There is no shortage of popular science books available in bookshops and libraries, but there are few analyses of the significance of this mode of representing science. Those that do tend to focus on the public understanding of science. These two general texts are particularly useful:

Broks, P. (2006) *Understanding popular science*, Maidenhead: Open University Press.

Gregory, J. and Miller, S. (1998) *Science in public: communication, culture, and credibility*, New York: Plenum.

Nelkin's classic work on science in the media is still well worth looking at:

Nelkin, D. (1995) *Selling science: how the press covers science and technology*, New York: W.H. Freeman.

Nelkin, D. and Lindee, M.S. (1995) *The DNA mystique: the gene as cultural icon*, New York: W.H. Freeman.

7

Science Fiction

It's too bad that one term – *science fiction* – has served for so many variants, and too bad also that this term has acquired a dubious if not downright sluttish reputation. True, the proliferation of sci-fi in the 1920s and 1930s gave rise to a great many bug-eyed-monster-bestrewn space operas, followed by films and television shows that drew heavily on this odiferous cache. . . .

In brilliant hands, however, the form can be brilliant.

Margaret Atwood, *In Other Worlds* (2011)

Science fiction (sf, SF, sci-fi) is a product of time and place, a creation of modernity: a society in which scientific knowledge and discourse were hidden or confined to a very small sphere of influence would not create these types of narratives. It is only in the nineteenth century that we see the emergence of speculative tales that deploy technology as a significant feature, and only in the late nineteenth century that we see speculative tales concerning people and societies of the future with a focus on science and technology (Alkon 2002). Much science fiction film, television and writing is specifically located in a particular time and place: to see this we need only think of the science fiction B-movies of the 1950s, the 'odoriferous cache' that Margaret Atwood refers to in the epigraph to this chapter, with their narratives describing invasions of evil aliens (which can be seen as metaphors for the Cold War threats faced by the USA (Seed 1999)), or the 1990s cyberpunk movement with its emphasis on the transformative power of the internet, personal computers and biotechnology (which was claimed to be presaging a transformation of identity and society (Cavallaro 2000)).

There is a paradox at the heart of science fiction, one that reflects a tension between science and the society it is a part of. 'Science fiction' is a contradictory term: our common-sense understanding of science is that it is a form of knowledge that describes what is the case whilst our societal understanding of fiction is that it describes what is not the case. Fiction and science do not sit easily with one another. However, as we shall see, much science fiction is reliant on themes and theories taken directly from contemporary scientific research, and some texts make an ostentatious display of their scientific and technical accuracy and credentials.

Science fiction is a popular topic for cultural theorists and cultural studies in general (Bould et al. 2011; Bukatman 1993; James and Mendlesohn 2003; Jancovich 1996; Johnston 2011; Kuhn 1990; Pearson et al. 2010; Penley 1989, 1997). For a genre that is firmly embedded in popular culture this is unsurprising: the discipline of cultural studies has long been concerned with popular, as opposed to high, culture and in particular with the emergence of new genres and new forms of expression that are specific to time and place. However, almost all cultural studies analysis of science fiction ignores an important facet of this genre: that some science fiction texts are about science itself, and all science fiction depicts, if only in part or only obliquely, social attitudes towards science, scientists and scientific institutions. This becomes particularly clear if we consider the role of gender in science fiction narratives. Mainstream science fiction writers are usually men (although not all, as we shall see), most of the main characters in these stories are men, and their plots often hinge on the tension and resolution between emotions (often represented by female characters) and rationality (embodied in male characters). Similarly, scientific activity in contemporary society is a largely male-dominated sphere where most leading scientists are men and the ideals of scientific endeavour are seen as the expression of pure rationality and the avoidance and rejection of subjective and emotional analyses. Our male-dominated science is represented by our male-dominated science fiction. The politics of representation visible here describe our androcentric science and culture. Science fiction, reliant as it is upon perceptions of the science that exists in our society, becomes a vehicle for the attitudes and experiences of people inside the institutions of science, and also those on the outside who form their opinions and construct their meaning of science from consuming cultural representations of science.

Many cultural studies of science fiction do consider in some depth what these narratives of futurity are saying about technology, and will identify some link between technology and scientific knowledge. Annette Kuhn's seminal collection of essays (Kuhn 1990) provides a range of perspectives on what science fiction cinema in contemporary society is about. The essays focus on, variously, the changing relationships between humans and non-humans; our use and understanding of technology; the transformation of humans by technology; the possible form future societies may take (and, by implication, the critique by science fiction of the present state of affairs); and the adoption by this genre of more widespread narratives such as that of eventual redemption, the monstrous feminine (the example used is the film *Alien*) and the primal scene fantasy. In addition, commentators focus on the audiences for science fiction narratives, considering how audiences use science fiction to understand and articulate their desires and construct identities (Penley 1997), or how the technology that is represented in science fiction is also the medium that we are using to access the representation. In these studies we can see that science fiction is, amongst other things, about technology, the self, identity, sexuality, politics and the environment.

These analytical approaches to the genre of science fiction – as well as noting that the concept of 'genre' itself is problematic – construct the text in different ways, focusing on different features and producing a wide range of results. However, what we don't see from such approaches is that science fiction is also about science. Cultural studies can identify a large number of facets of science fiction narratives, but rarely focuses on one of the core sets of beliefs that underpin science fiction narratives, namely 'faith' in the power of science. This is by no means a failing on the part of cultural studies: examinations of science fiction narratives as indicative of unconscious desires, as expressions of gender relations, as examples of latent or hidden orientalism, are all important outcomes of cultural analysis. Using an avowedly popular culture genre such as science fiction to show that our society is actively involved in constructing others and ordering social divisions through cultural production is a profoundly important task, and it is not the aim of this chapter to detract from this in any way. Yet it remains the case that such approaches avoid discussion of what we are learning from science fiction about science and its role in our society.

In the rest of this chapter we will look at what science fiction tells us about science. At a very general level science fiction narratives reinforce some of our societal attitudes towards science (although this happens in conflicting ways), and at a specific level there are some subgenres of science fiction that deliberately locate themselves inside a scientific worldview and celebrate science, scientists and scientific knowledge. Overall this chapter will argue that science fiction becomes a resource in the social construction of shared ideas about science, but representations of science are also a consequence of shared meanings and understandings of science in society. There is a dynamic relationship between the cultural productions and the social meanings attached to knowledge, people and institutions, in this case science. Because of this dynamic and changing relationship it is difficult to provide hard and fast definitions of the phenomena we are looking at, so we will start by trying to place some boundaries around these. A starting point is to ask 'what is science fiction?'

What is science fiction?

Almost all cultural studies analyses of science fiction commence with a similar formulation: that it is not possible to provide a comprehensive definition of what science fiction is. For example, the editors' introduction to *The Routledge Companion to Science Fiction* states: 'Therefore, one principal aim of [this book] is to bring into dialogue some of the many perspectives on the genre, without striving to resolve this multiplicity into a single image of sf or a single story of its history and meaning' (Bould et al. 2011: xx). Their wariness is sensible, given the range of things that we call 'science fiction', but it is worth noting that most people who are familiar with contemporary cultural production will have little trouble in identifying what they consider to be science fiction. They will have a good idea of what it is that makes something into 'science fiction'. The process of definition itself is significant: why are cultural productions placed in genres? The ease with which science fiction texts in contemporary culture can be identified is no doubt facilitated by the prevalence of categorized cultural productions, and we must be aware that such categorizations are made, at least in part, by the organizations that have a vested interest in ensuring maximum profits through maximizing audiences, or through constructing new, niche, markets. Similarly, audiences and authors may deliberately choose to locate themselves inside a genre category for reasons of identity construction or community formation.

Keith Johnston's *Science Fiction Film: A Critical Introduction* (2011) also starts from this point of uncertainty, but goes on to provide a fairly comprehensive list of what science fiction film is:

> a popular fictional genre that engages with (and visualizes) cultural debates around one or more of the following: the future, artificial creation, technological invention, extraterrestrial contact, time travel, physical or mental mutation, scientific experimentation, or fantastic natural disaster. Science fiction films are traditionally dramas about these topics, usually with thrilling and romantic elements and often reliant upon state-of-the-art special effects techniques to create a new, or expanded, worldview. (ibid.: 1)

Most science fiction fans would probably agree with this definition, and Johnston is certainly correct in pointing out just how popular the genre as a whole is. Science fiction films are major income earners and attract large audiences, and science fiction is, perhaps, the most widely distributed fictional genre in terms of the media it appears in: films, television, novels are obvious, but science fiction dominates comic books and graphic novels, appears in contemporary popular music, contemporary art (Paolozzi's work often referenced robots and spaceships), architecture, graphic design and advertising. We are deluged with science fiction texts in contemporary culture, and will often reach for science fiction themes to help us explain the world: the use of the term 'Frankenstein foods' or 'Frankenfoods' to describe GM crops is taken directly from Mary Shelley's science-inflected horror story *Frankenstein* (Shelley [1818] 1994), and its promulgation is widely considered to be one reason why GM foods have failed to be accepted in UK society.

Despite, or perhaps because of, this popularity, science fiction is often seen as being a form of fiction that is not as good or as legitimate as other forms of fictional writing. Whilst this is almost certainly changing with time, allowing science fiction a much more central role in popular culture today, science fiction was a marginalized genre whose position gave it a much greater degree of latitude in terms of what it could include. Cultural theorists recognize this:

> One needs to turn to 'minor', not to say marginal and hybrid genres, such as science fiction, science fiction horror and cyber punk, to find fitting cultural illustrations of the changes and transformations that are taking place in the forms of relations available in our post-human present. Low cultural genres, like science fiction, are mercifully free of grandiose pretensions – of the aesthetic or cognitive kind – and thus end up being a more accurate and honest depiction of contemporary culture than other, more self-consciously 'representational' genres. The quest for positive social and cultural representations of hybrid, monstrous, abject and alien others in such a way as to subvert the construction and consumption of pejorative differences, makes the science fiction genre an ideal breeding ground to explore our relation to what Haraway describes affectionately as 'the promises of monsters'. (Braidotti 2006: 203)

Science fiction, by this account, far from being fantastical and unrealistic, is one of the few vehicles that can really address how we are changing as humans as we enter the future. And this has always been the case in science fiction: for example, the first on-screen TV kiss between a black and a white character took place in an episode of the 1960s TV science fiction *Star Trek*, a programme that presents a fantasy future.

The producers of science fiction texts seem to have fewer problems in constructing definitions of what they are doing, and they often return to common themes of the future, space and technology:

> Now – contrary to a general belief – prediction is not the main purpose of science-fiction writers; few, if any, have ever claimed 'this is how it will be'. Most of them are concerned with the play of ideas, and the exploration of novel concepts in science and discovery. 'What if . . . ?' is the thought underlying all writing in this field. What if a man could become invisible? What if we could travel into the future? What if there is intelligent life elsewhere in the Universe? These are the initial grains around

which the writer secretes his modest pearl. No one is more surprised than he is, if it turns out that he has indeed forecast the pattern of future events. (Clarke 1977: 3)

Arthur C. Clarke (1917–2008) was one of the most influential science fiction authors of the twentieth century, and with a career spanning seven decades his work is an interesting 'barometer' of the course of science fiction writing. Clarke's definition of science fiction deals with speculation about possible futures, and largely his science fiction writing stuck to that. Science fiction, by this definition, can be about just about anything, as long as it is looking at the future and answering a 'what if?' question. But this is not a sufficient working definition of science fiction: clearly, asking and answering a 'what if?' question of the form 'what if a woman were to fall in love with her best friend's partner?' could be the start of a romance, a science fiction, a murder mystery, a screwball comedy, etc. '[W]hilst SF is imaginative fiction, it does not follow that all imaginative fiction can be usefully categorized as SF' (Roberts 2006: 4). We need additional components to discriminate what science fiction is:

Years ago I was working in Schenectady for General Electric, completely surrounded by machines and ideas for machines, so I wrote a novel about people and machines, and machines frequently got the best of it, as machines will. (It was called *Player Piano*, and it was brought out again in both hard cover and paperback.) And I learned from the reviewers that I was a science-fiction writer. I didn't know that. I supposed I was writing a novel about life, about things I could not avoid seeing and hearing in Schenectady, a very real town, awkwardly set in the gruesome now. I have been a soreheaded occupant of a file drawer labelled 'science fiction' ever since, and I would like out, particularly since so many serious critics regularly mistake the drawer for a urinal. The way a person gets into this drawer, apparently, is to notice technology. The feeling persists that no one can simultaneously be a respectable writer and understand how a refrigerator works, just as no gentleman wears a brown suit in the city. (Vonnegut 1976: 25)

Kurt Vonnegut (1922–2007) identifies the second point of our starting definition of science fiction: it is about technology and, by inference – as technology is seen as a product of science – science. But Vonnegut also reminds us of the 'lower' status of science fiction as a genre, at least at the time of his writing: indeed, a central character in many of Vonnegut's books is Kilgore Trout, a failed but amazingly prolific science fiction writer, whose works – which appear as filler in cheap pornographic magazines and are largely unread – contain the answers to most of life's mysteries. Margaret Atwood, quoted in the epigraph to this chapter, is also a reluctant inhabitant of that file drawer, although perhaps less so than Vonnegut. Atwood chooses the term 'speculative fiction' to describe her work.

These three quotations from prominent science fiction authors provide us with a good starting-point definition: science fiction is about a future (not *the* future), is about largely plausible possibilities and has a focus on technology. Most science fiction will conform to this definition. However, science fiction, like science, is a family resemblance concept that has blurred edges, a great many subdivisions and a wide range of meanings attached to it by different audiences. Clearly a film such as *Avatar* (James Cameron, 2009) is science fiction: it is about the future and our relationship to technology (robots and computers), and presents humans in conflict with technological forces. How about George Orwell's *Nineteen Eighty-Four* (1954)? It is about the future (or at least it was when it was written), has a strong focus on technology (telescreens, memory holes, etc.) and presents humans in conflict with technological forces (the attempts of Julia and Winston to escape their constant surveillance). *Avatar* has a 'happy' ending with the Na'vi, indigenous creatures of the planet Pandora, being saved from destruction due to the humanity of a few gallant heroes and scientists; *Nineteen Eighty-Four* has an unhappy ending with the humanity of the heroes being destroyed by the technocracy and dictatorship that surround them. Would you describe *Nineteen Eighty-Four* as science fiction? The context the text emerges in clearly plays some role in how we define it. I bought two copies of Orwell's book in 1979: my British copy suggests it should be filed under 'fiction literature', my Yugoslav copy clearly categorized it as 'naučana fantastika' – science fiction. The novel based on James Cameron's screenplay of *Avatar* (Harper 2009) will be located in the science fiction section of your local bookshop, but Orwell's *Nineteen Eighty-Four* will be in the literature section of the same shop. You may have been encouraged to read *Nineteen Eighty-Four* as part of an English literature course at school, but *Avatar* was probably not on the same syllabus. Yet *Nineteen Eighty-Four* displays the characteristics we look for and closely identify with science fiction: future orientation and a focus on new forms of technology. These two characteristics will be clearly visible in most science fiction that we consider: we could, of course, add a huge number of other common themes (such as space exploration, or contact with alien species), but not all science fiction texts will contain these, and it is often science fiction by women writers that departs from these standard themes. All will, however, look to the future and make technology a key issue to be utilized in the text, and we can use this as a starting-point definition. (And yet even with such a broad definition we will find exceptions. For example, William Gibson and Bruce Sterling's *The Difference Engine* (Gibson and Sterling 1991) is an examination of a fictional mid-Victorian society that has been transformed by the introduction of information technology. Clearly science fiction, but set in the past not the future, this was one of the first major 'steampunk' novels, a subgenre of science fiction that became popular in the late 1980s and throughout the 1990s. The best-known steampunk production was probably the 2003 film adaptation, directed by Stephen Norrington, of Alan Moore and Kevin O'Neill's comic book *The League of Extraordinary Gentlemen* (Moore and O'Neill 2000).)

Problems in producing hard and fast definitions of science fiction notwithstanding, we can look at a range of science fiction texts to provide examples of our relationship to science and technology in contemporary society, and instances of the articulation of the discourse of scientism. The main focus here will be on what mainstream science fiction, and subsequently what a number of subgenres of science fiction, tell us about science in society. We will use science fiction representations of science as a means of understanding general societal concerns about science, and general societal discourses of science, and we will start by taking a trip to the movies.

Mainstream science fiction – definitions and its relation to science

Notwithstanding the difficulties of defining science fiction and the problems associated with analysing genres, there is something we can identify as being 'mainstream' science fiction, that is, science fictions that receive widespread public attention through their promotion in a range of media. Again, boundaries surrounding this definition will be fuzzy: the *Star Trek* TV series could clearly be defined as mainstream science fiction, as could *Doctor Who*: both have extensive exposure, and would be clearly defined as science fiction by audiences (although it is notable that both of these texts have significant soap opera elements). However, we will encounter more difficulties in categorizing a film such as *2001: A Space Odyssey*: certainly one of the most famous science fiction films of the twentieth century, and arguably one of the best – but mainstream? Stanley Kubrick's 1968 film, with its strict realism, hard-to-follow story and often silent soundtrack (considered by many to be the most scientifically accurate sci-fi film ever made (Kirby 2013: 1)), diverges considerably from what is expected in a science fiction film, yet it is frequently shown on television, and in the year 2001 was re-released to nationwide cinemas in the UK. Suffice to say that the construction of a category such as 'mainstream' science fiction would be problematical if we were to try and sustain a form of categorization in a systematic way. That is not the intention here: rather, I want us to be able to distinguish science fiction that a large part of the general public will have an opportunity to encounter, on the one hand, from more marginal and specialized forms of science fiction, on the other. In the case of science fiction films, we will look at some of the blockbusters released in 2013 that had a science fiction component: *Gravity* (Alfonso Cuarón, 2013), *Despicable Me 2* (Pierre Coffin and Chris Renaud, 2013), *World War Z* (Marc Forster, 2013) and *Star Trek Into Darkness* (J.J. Abrams, 2013).

Mainstream science fiction films

Do films such as *Gravity, Despicable Me 2, World War Z* (WWZ) and *Star Trek Into Darkness* tell us anything about science? All four feature technology as a significant feature – although in different ways; two of the four (*WWZ* and *Despicable Me 2*) feature science as an important part of the plot; one (*Star Trek Into Darkness*) is set in the far future – the others are set in alternative presents. In contemporary society scientism tells us a story of our technoscientific present and future: that science will generate new technologies that will change the world and, by extension, ourselves. Aspects of these films may, directly or indirectly, reinforce our general societal understanding of science.

Gravity was an international blockbuster in 2013, and received seven Oscars including best director for Alfonso Cuarón. Set in a fictional recent past, the film portrays a near-Earth space disaster that leaves Sandra Bullock and George Clooney, two NASA astronauts on the space shuttle (STS157), fighting to survive in the hostile environment of space. The film relies on very realistic sets which present a realist, almost hyper-realist, visualization of the near-Earth space environment and the human-made structures it contains. This was done quite deliberately to give this science fiction film a non-science fiction look:

> We always thought it had to look like one of those NASA documentaries. We didn't want it to look like science fiction; we didn't want it to look like a comic book – we wanted it to look like a documentary. (Alfonso Cuarón, interview, http://www.deadline.com/2013/12/oscars-alfonso-cuaron-gravity-interview)

Why would a science fiction film director want their film *not* to look like science fiction? There are a number of reasons, the most obvious being that if the film can break out of its genre category it can, perhaps, attract a wider audience. In addition, making the film look like a documentary gives the narrative a different status and the audience may respond differently. Finally, *Gravity*, as well as being a thrilling space adventure, is also an extended metaphor for grief and loss and resolution as the lead character, played by Sandra Bullock, grieving the loss of her daughter, then loses almost all of the material objects that sustain her and faces almost certain death herself. The film finishes with a classic 'rebirthing' sequence as Bullock swims then crawls out of the sea. The film thus references both the very highest technology – spaceflight – and the dawn of humankind (emerging from the primordial sea). There are clear references to *2001: A Space Odyssey*, not only in terms of trying to be honest about space environments, but also through reproducing some of the classic images in the 1968 film: the scene with Bullock weightless in the International Space Station airlock, floating with the air tube of her spacesuit looking like an umbilical cord, is an homage to the 'spacechild' image of *2001* with the human foetus floating against a circular background.

Cuarón: 'We tried to be as accurate as we could within the framework of our fiction. In the end, it's fiction and it's an emotional journey more than anything else.' (quoted in Pulver 2013)

The film is a strong reminder that spaceflight is a remarkable technological achievement and that technology is vital to how we live our lives, even our very survival. It relies on its audience understanding the achievement of spaceflight and having a sense of awe and wonder at what technology can do in helping us to understand our place in the universe.

Star Trek has been a cult phenomenon since it started, and the films, and TV programmes, attract huge audiences and very loyal fans. *Star Trek* has been influential both within science fiction and beyond (Penley 1997). Set in 2259 in a galaxy populated by humans and aliens, the twelfth instalment in the film series, *Star Trek Into Darkness*, carries on the original franchise using the same cast of characters that populated the science fiction TV series from 1966. As with the other films and TV programmes the lead character, Captain Kirk of the Starship *Enterprise*, and his loyal crew face alien spaceships, exploding planets and almost insurmountable odds, but triumph in the end over the forces of evil arrayed against them. In a very 'science fiction'-style plot twist, the baddie in this film turns out to be Khan, a character from the original first TV series of *Star Trek* who also returned in the 1982 film *Star Trek II: The Wrath of Khan* (Nicholas Meyer, 1982).

The film is a good example of the subgenre of space opera (see the later subsection on this), where heroic spaceship captains save the world after monumental and visually exciting space battles. Films such as this remind the audience that human technology will advance inexorably; they often make explicit reference to earlier human triumphs in flight and spaceflight, and there is a relentless expansion of both technology and humans into the future and the galaxy. The societies of the future are shaped by the technologies that are available to them, although it is often the case in space opera that whilst technologies and social forms change because of technological development, human character does not, allowing the deployment of very old and easily recognized narrative tropes. These films and stories articulate a strong form of technological determinism (see box 7.1), where the technology engendered by scientific knowledge takes on a momentum that is hard or impossible to stop.

Box 7.1 Technological determinism

We can see technological determinism in many explanations of the world. Why do children get obese? Because they spend their time in front of computers rather than running around; thus computers have changed the world around us. Why is our society 'superior' to other societies? Because we have higher levels of technological development. How do we explain the rise of our industrial society in history? Through the innovations brought about in the late eighteenth century, notably the steam engine and other technologies that allowed industrialization to take place.

Technology of course does have significant effects on our society: just think about how our society has been transformed by the invention of antibiotics, contraception and anaesthesia. However, there is a fundamental problem with technological deterministic explanations, and this problem connects very strongly to scientism. It is that technological determinism cannot explain or account for technological change itself: it shows technologies as emerging seamlessly through some hidden logic of their own, or through being a by-product of science, which in turn is then represented as having an inexorable, and hidden, inner logic of emergence. Technological determinism hides all human social contributions to innovation, and, equally significantly, gives us an often inaccurate understanding of what technology means to us. Finally, the idea of a relentless logic of technological innovation, with discovery piling on top of discovery and social change resulting from this, is wrong in a large number of specific cases. Technologies are 'pulled' into being by a combination of things: construction of markets, articulation of social needs and deliberate R&D into areas perceived by states or corporations as needing investment. They don't emerge because of their own inner 'logic'.

Despicable Me 2 (a 3D computer-animated comedy film) starts with a super-secret scientific base being stolen from the Arctic Circle. The scientists there had been working on PX-41, a powerful mutagen that turns harmless creatures into savage purple monsters. Gru, a former super-villain, is recruited by the Anti Villain League (AVL) to find the super-villain who stole the PX-41; the suspect is believed to be working in a shopping mall, so Gru and his AVL partner Lucy go undercover. They find the super-villain El Macho, working with Dr Nefario, Gru's former scientist and inventor; the PX-41 has been applied to Gru's minions and they will be sent by rocket around the planet, thus giving El Macho world domination. Lucy is captured, Dr Nefario changes sides and develops an antidote to the PX-41, and Gru and Dr Nefario overcome El Macho with a fart gun and a lipstick taser. Gru rescues Lucy and they get married. What does the film say to its (young) audience?

This is a very high-tech alternative reality, although in terms of some of the gadgets being deployed not so far from the kind of technology used by James Bond or other super-spies in films aimed at older audiences. The technology is powerful, very effective and efficient; in almost all cases it works exactly as it is meant to, and Gru can ask Dr Nefario to make just about anything he wants. The scientists working on the PX-41 mutagen are presented in stereotypical ways: secretive, working for an anonymous government institution, wearing white coats, surrounded by recognizable lab equipment, all male and, perhaps most significantly, carrying out experiments on animals (cute bunnies). This is a version of science as separate, remote, cold (the lab is, after all, located in the Arctic Circle), powerful and without morals.

World War Z is a disaster horror film set in the present day. A zombie pandemic threatens all of humanity and the United Nations send their best investigator, Gerry Lane, accompanied by a top virologist, Dr Andrew Fassbach, to discover the cause and possible remedy. The first two thirds of the film are high-action zombie-fighting scenes; the final third is set in a World Health Organization (WHO) laboratory near Cardiff where Lane proposes a theory to defeat the zombies: injecting a healthy human with a pathogen acts as a form of camouflage against the zombies. The WHO scientists agree to help Lane retrieve the necessary pathogens from the zombie-infested labs. Unlike most other zombie films, where the plot is based on besieged humans trying to survive a zombie onslaught, *WWZ* is a chase film with the heroes desperately trying to find a cure or vaccine for the pandemic. The representation of scientists is interesting. The virologist who is the UN's great hope for finding the cure is hubristic and self-assured, certain that his scientific and analytical approach will bring a successful result. He is stereotypical in two ways: the scientist's sense of certainty and power over nature is a trope we see repeatedly in modern culture, and the idea that a lone (male) genius scientist will fix the problem is also a recurring trope. Later in the film, the WHO scientists are portrayed as dedicated and concerned, but the film alludes to some secret and dangerous activities that have been taking place at this unexplained WHO facility in rural Wales. Again, the trope of secretive, remote and cold science is being deployed.

To advise on how the zombies should behave, the film's producers hired in a science consultant, David Hughes – an evolutionary biologist at Pennsylvania State University. Hughes is an expert on 'zombie ants' – ants infected with a parasitic fungus that takes control of the insects' brains:

> Hughes explained that one way 'World War Z' zombies are different than zombies seen in other films is that these zombies don't compete with each other. 'If everyone was infected with the same virus, they would exhibit collective behavior,' said Hughes. 'Using the idea of swarm intelligence, the zombies would work together and overcome obstacles they could not alone, just as ants do.' (Lorditch 2013)

The film doesn't pretend to be 'scientifically realistic'; human zombies don't, and can't, exist, and perpetuating a myth about the possibility of zombies could be seen as being pseudoscientific. So why did Hughes agree to take part? 'I have been able to bring biology and evolution to a different audience which leads to conversations. ... People are generally good biologists because we observe it [the science of biology] from the cradle to the grave. So why not use Hollywood to get the message across?' (ibid.). Scientists' involvement in films is clearly a risky business: the opportunity to shape the science content of the film, and to have some fun, are clear benefits. 'Involvement can also help those shows to portray scientists as real people and role models rather than as negative or laughable stereotype' (Smaglik 2014: 113). But there are risks and it is unlikely that the collaboration will be even handed: film-makers will have priorities that they want to achieve and scientific accuracy will almost certainly not be the main one.

What do the film-makers get from collaborations with scientists? David Kirby's extensive analyses of collaborations between scientists and film-makers, focusing on 'science consultants, whose advice shapes the narrative and visual content of specific cinematic texts' (2013: xiii) and which are not confined to just science fiction films, found that:

> The motivation for filmmakers to utilize science consultants is clear. Scientific knowledge holds a place of privilege in society, and the scientific expert is often used to legitimate one's own views. By using scientists as consultants, filmmakers can claim legitimacy for their visions of science. The publicity value of science consultants is evident in the fact that filmmakers frequently highlight scientists in a film's press and marketing material. Studios encourage scientists to speak with the press about their film work; scientists also often attend press conferences surrounding the films. (Kirby 2003: 264)

Kirby's research suggests a different outlook will help to make sense of what we are really looking for. There is a generally held idea that the public needs to be educated, to gain scientific literacy, about what science really is, and that their ignorance can only be countered by providing scientific knowledge; we saw this deficit model quite clearly in our analysis of popular science texts. However, Kirby suggests that applying the deficit model (see box 7.2) to *fiction* is unhelpful, and the scientific community's concern for 'accuracy' in fiction is misplaced. Firstly, having scientists involved can help to make representations of scientific research more realistic, whilst not being slaves to accuracy. Second, in what way would we expect scientists represented in films by actors to be 'accurate'? Is there some way of portraying a 'realistic' scientist? Almost certainly not, but science consultants can help film-makers convey the sense of wonder and excitement in scientific research, and help them avoid tired stereotypes:

> While increasing 'scientific accuracy' in fiction may not enhance the public understanding of science as prescribed by the deficit model, the presence of scientists in the filmmaking process can improve the public understanding of science; that is, if we take scientists' concern with 'public understanding of science' to mean more than public 'appreciation' of science. (Kirby 2003: 274)

Science consultants need particular qualities, not least of which is recognizing that they are not the most important person on the set (unlike, perhaps, their position in their day-job laboratory) and that they cannot ensure scientific accuracy throughout: one of the advisors on *Gravity* commented that his role requires soft skills and a thick skin; scientists are likely to be respected if they are not obstructionists. But science consultants' rewards are often meagre: they don't often get to meet the celebrities, rarely get paid (although other rewards, such as having your formula written on a whiteboard in the background of a scene, or your textbook being prominently displayed, may bring some kudos) and may not be listened to. Kirby says he gets an email a month from a science researcher wanting to break into the films: 'I kind of feel bad telling them they probably can't make a living doing this' (quoted in Smaglik 2014: 113).

Box 7.2 PUS, PEST or CUSP?

Steve Fuller described the public understanding of science (PUS) as 'our latest moral panic' (Fuller 1997); the analysis of the discourse of popular science in chapter 6 shows that there is still much consternation about the lay public being 'ignorant' of science. But this argument is based on a 'deficit model' – the idea that the public simply need to have more science poured into them to alleviate this deficit. The deficit model of the public understanding of science is a red herring, one that serves to reinforce the status of science as a superior form of knowledge, but also one that is used to excuse the scientific community for poor communication of its work by blaming non-scientists for a lack of understanding. A more robust and appropriate idea of public engagement with science and technology (PEST) has been adopted more recently to try and investigate the relationship between science and its publics, and to try and foster a more mutually aware relationship. This has been reflected, at least in part, in UK science research councils' insistence that publicly funded scientific research should be disseminated to the public; after all, it is the public who are paying for this. But this is not much of a two-way street, and much dissemination is done in a top-down way, rather than inviting the public into the sites of scientific research (Toumey 2006). Perhaps what is needed is CUSP – critical understanding of science in public – whereby the public are involved in, and aware of, the negotiation and construction of the meaning of science in public (Broks 2006: 94).

Three recurring themes – knowledge leads to technology that will change our world; scientists are separate, cold and emotionless; scientific institutions are part of the structure of power in society – are common to a great many mainstream science fiction narratives that describe contemporary society or near-future alternatives. Science, scientists and scientific institutions are not the only theme of the films, but they do present a consistent and fairly coherent image of science as a difficult thing that is hard to understand and has inherent power, an articulation of the discourse of scientism. Mainstream science fiction films rely not only upon general societal attitudes and cultural representations of science for their effect and impact, but also for the audience to position the film.

Science fiction trends and tendencies

Science fiction, even 'mainstream' science fiction, encompasses a huge range of different types of narratives, appears in a range of media and appeals to a diverse audience. There are a large number of subgenres in science fiction – some obvious and self-proclaimed (such as steampunk, mentioned earlier), others less clear. Some subgenres come and go; cyberpunk, for example, is a hybrid genre bringing together hard-boiled detective fiction with near-future dystopian scenarios that often feature artificial intelligences, computer hackers and cybernetics. The first cyberpunk novel to be recognized as such was William Gibson's *Neuromancer* (1984), and the first cyberpunk film was *Blade Runner*

(Ridley Scott, 1982). Both are considered to be science fiction 'classics', but the cyberpunk movement of the 1990s seems to be on the wane now. Other subgenres have risen up, but as with the definitional problem of science fiction itself, it is difficult to provide hard and fast definitions of these, or even an agreed list of which subgenres are extant at present.

The Cambridge Companion to Science Fiction (James and Mendlesohn 2003) lists four notable subgenres: hard science fiction, space opera, alternate histories, and utopias and anti-utopias. *The Routledge Companion to Science Fiction* (Bould et al. 2011) lists twelve: alternate history, apocalyptic sf, arthouse sf film, blockbuster sf film, dystopia, eutopia, feminist sf, future history, hard sf, slipstream, space opera and weird fiction. Of course, there will be crossovers and mash-ups between these subgenres, and between them and other genres (including those not traditionally considered to be science fiction). It is significant that neither list includes 'fantasy', another popular subgenre. The exclusion of fantasy is quite deliberate: unlike the genres listed, fantasy ignores science and replaces it with some version of magic, spirituality or religion as a mechanism for bringing about change. It retains a connection to science fiction in general by, often, locating its narratives in the future and in alternative worlds. In contrast, the books we focus on in what follows all use science, to varying degrees, as a way of explaining the world as depicted in the text. But we should note that fantasy–science fiction crossovers also exist: Peter F. Hamilton's *Night's Dawn* trilogy presented a space opera–fantasy crossover whose plotline centres on dead souls returning from some sort of pre-heaven limbo to inhabit the bodies of the living (Hamilton 1996, 1997, 1999).

Once again we will face problems in closing boundaries around our definitions. In a number of cases, science fiction texts will quite easily fall into more than one category, and others will not admit to categorization at all. As with our discussion of mainstream science fiction, the aim here is not to produce rigid and systematic classifications; rather it is to locate trends and tendencies in this area of cultural production. Here we will focus on two examples from science fiction subgenres that are mentioned in both of those lists: space opera and hard science fiction.

Space opera

'Space opera is the most common, and least respected, form of science fiction' (Westfahl 2003: 197). Paraphrasing Wilson Tucker, who coined the term in the 1940s, Westfahl identifies three key features that a space opera must have: a spaceship, an exciting adventure story and a tendency towards a formulaic plot or mediocrity. Andy Sawyer, writing on the same topic, notes that space opera is, 'essentially, a mode for adolescents, particularly American adolescents' (Sawyer 2011: 506).

These are harsh judgements on what is, after all, the most popular subgenre in science fiction, the most easily recognized by non-science-fiction audiences, and the most fun to read (for me, anyway). Sawyer does go on to recognize that the alleged pointlessness of space opera is 'debatable' (ibid.: 506) and that its definition is also contestable:

> Not all fictions involving space travel, conflict, huge vistas, and Big Dumb Objects are space opera, which is committed to action and adventure, focused upon the heroic, and frequently takes a series or serial form which allows for either a sense of escalation or constant variations on a comparatively narrow set of themes. (Sawyer 2011: 505)

The roots of space opera go back to, at least, the 1930s when Hollywood produced serials such as *Flash Gordon*. In terms of books, the most important early space opera was Isaac Asimov's *Foundation* trilogy, written in the 1950s and subsequently expanded in the 1970s and 1980s (Asimov 1979a, 1979b, 1982, 1983a, 1983b, 1985, 1986, 1988, 1993): the sequential aspect of Asimov's work was new and was subsequently copied by many other science fiction writers. In television, the clearest example is the *Star Trek* series of the late 1960s, again expanded and extended in the 1980s, 1990s and 2000s with the sequel series *Star Trek: The Next Generation* and spin-off series *Deep Space Nine*. Space opera films are legion, typified by the *Star Wars* and *Star Trek* series of films. All share the characteristics of intergalactic travel, encounters with aliens and conflict-based plots. In recent years, however, there has been a subgenre emerging in space opera – postmodern space opera – which departs from the traditional theme of good versus evil coupled to romance and tackles more moral and political issues, whilst retaining the battling spaceships and Big Dumb Objects motifs. Iain M. Banks's *Culture* series of novels is the best-known example of this trend. Whilst space operas generally avoid in-depth discussions of quite how faster-than-light travel and teleportation could possibly happen, they do connect to science in a range of ways. Here we will focus on one author's space opera to illustrate key features of the uptake and articulation of a vision of science by science fiction.

Ann Leckie's novel *Ancillary Justice* (2013) clearly fits the parameters of space opera. It is set in the far future in a galactic empire that is patrolled by intelligent and incredibly powerful starships. It has an exciting plotline: will ancillary Breq be able to exact revenge on the ruler of the galactic empire who callously murdered her entire crew, as well as countless innocent civilians? It includes elements of romance: Seivarden, Breq's companion, gradually falls in love with Breq, but this love is, perhaps, unrequited. There are also alien species, incredible technologies and moral questions peppered throughout the text. However, Leckie spends very little time explaining either the back story of how such a galactic empire came into being (we don't, for example, even know if it is an outcome of human expansion from planet Earth or is a wholly separate phenomenon such as the *Star Wars* 'universe') or how the technologies that allow interstellar travel and awesome firepower are possible. As well as being an exciting piece of science fiction, Leckie's book can also be read as a kind of critique of contemporary geopolitics. Ancillary Breq (who is an artificial intelligence) is enslaved to obey the orders of Anaander Mianaai, the supreme leader of the Radch empire, but realizes that she is being used in ways that are immoral. Her struggle with herself and then with Anaander Mianaai mirrors the ethical choices that confront those who serve political institutions that carry out immoral military actions. Written at a time when the US military were withdrawing from foreign conflicts in Afghanistan and Iraq, *Ancillary Justice* reads as a strong critique of using military personnel to achieve by force what should be addressed by political, or other, means.

Science features in *Ancillary Justice* as a sub-current to the technology that is deployed throughout the book. The Radch empire has achieved its pre-eminence not only through having better military technology, but also through using science to produce ways of ensuring compliant individuals and populations through 'reeducation'. But human want has been eradicated through having better technologies: there is no hunger or disease, although there is still considerable inequality, in Radch society. Leckie's faith in technology is a form of faith in science, a neutral force that can be harnessed for good or evil.

Hard science fiction

Some science fiction texts deliberately embrace and promote the 'science' in their fiction. '[A] work of sf is hard sf if a relationship to and knowledge of science and technology is central to the work. . . . The primary characteristic for defining a work as hard sf is its relationship with science' (Cramer 2003: 187–8). The subgenre 'hard science fiction' is, arguably, the core around which all of science fiction revolves (Samuelson 2011: 494) and has been for a long time. Some critics will identify the writings of Jules Verne or Edgar Allan Poe as having characteristics of hard science fiction; we can trace hard science fiction in a fairly continuous line back to the early phase of the popularization of science fiction in the 1940s and 1950s when it was articulated by writers such as Isaac Asimov and Arthur C. Clarke. Hard science fiction texts will often incorporate quite extensive discussions of scientific research and scientific principles: as such, it is unsurprising that few mainstream films or TV programmes are part of this subgenre, although a notable exception is Stanley Kubrick and Arthur C. Clarke's *2001: A Space Odyssey* (book and film 1968). The relationship of hard sf and science is reaffirmed every week by the most-cited international science journal *Nature*, which prints an sf story on the inside back page, often one written by a formal scientist.

Hard science fiction attempts to be accurate according to scientific principles: it doesn't 'take liberties' with the laws of physics (e.g. it will take seriously Einstein's theory of relativity, which posits that faster-than-light space travel is not possible), extrapolates from current trends in scientific research (often in consultation with scientific researchers) and tries to suggest genuine possibilities that may appear in the future. 'All sf requires some relationship to modern science and technology, including expectations of technological change. Hard sf ties much of its credibility to scientific rules and probability, enough that some readers define it simply as "hard" to read' (Samuelson 2011: 494). Hard science fiction is often associated with 'hard' science disciplines such as physics, astrophysics and astronomy, and some writers will quite deliberately display their formal scientific credentials to make this association clear. Alastair Reynolds's dust-jacket biography on his novel *On the Steel Breeze* (2013) tells readers that '[h]e gained a PhD in astronomy and worked as an astrophysicist for the European Space Agency before becoming a full-time writer'. Stephen Baxter, a prolific writer who does not confine himself to science fiction, but whose science fiction tends towards 'hard' and who collaborated with Arthur C. Clarke (himself a former government scientist and science advisor) on two hard science fiction novels, presents his credentials on his website:

> I have degrees in mathematics, from Cambridge University, engineering, from Southampton University, and in business administration, from Henley Management College. I worked as a teacher of maths and physics, and for several years in information technology. I am a Chartered Engineer and Fellow of the British Interplanetary Society. (http://www.stephen-baxter.com/author.html)

Baxter's novel *Proxima* (2013) is about humans travelling to and colonizing an 'exoplanet', i.e. a planet in another solar system. In the afterword to the book Baxter presents a list of formal science sources that he consulted in producing the novel, thus grounding the speculative story in science 'fact'. The realism of hard science fiction is an echo of the realism that is central to the standard account of science, and the realism at the heart of scientism – that science provides truthful accounts of the natural world; this type of narrative makes similar claims to the truth of the facts that it deploys as narrative devices to explain the plot. 'Underpinned by facts and theory . . . hard sf reflects scientific interpretations of reality' (Samuelson 2011: 494–5).

These texts share an attitude towards science, and an affection for science, that will be clear to the audience. Frequently the heroes of hard science fiction texts will be scientists: a mathematician (Hari Seldon in Asimov's *Foundation* series), biologists and computer programmers (Clarke's family of heroes in the *Rama* cycle of novels (Clarke 1974; Clarke and Lee 1991, 1992, 1994)), astronautical engineers (Baxter's Gary Boyle in *Flood* (2008) and Holle Groundwater in *Flood*'s sequel *Ark* (2010)) or a naturalist (Alastair Reynolds's Geoffrey Akinya in *Blue Remembered Earth* (2012) and *On the Steel Breeze* (2013)). The tasks and adventures that they become involved in often require the application of technical skills and scientific knowledge rather than force, emotion or cunning to ensure success. In contrast, the lead characters in space operas are often starship captains, soldiers or mercenaries (e.g. Hamilton's heroes in the *Void* trilogy (Hamilton 2007, 2008, 2010), Iain M. Banks's characters in his *Culture* novels (Banks 1987, 1988, 1990, 1991, 1996, 1998, 2000, 2004, 2008, 2010, 2012)), and the main protagonists of cyberpunk novels are computer hackers, streetwise cycle couriers, assassins and culture industry stars. In hard science fiction novels, having technical knowledge is often as important as more traditional heroic qualities. However, other 'traditional' heroic aspects are visible in the main characters in these hard science fiction novels: they are often men, and their expression of scientific rationality as the best way of coping with problems and crises is a strong validation of the 'standard' account of science.

Although much hard science fiction is about space travel, this is not the defining feature of the subgenre. Being 'true to' and making science and technology a major component of the narrative can bring in different formal science disciplines. For example, Madeline Ashby's *vN* (2012) describes a near future where robots (von Neumann machines – hence vN) are common in human societies. The main character, Amy, is a vN who is on the run after cannibalizing her grandmother, who was attacking humans, something that the vN robots' failsafe device should prevent. Amy and her vN companion are captured by human bounty hunters Rick and Melissa:

> Rick said: 'I'd watch out if I were you, buddy. She's a zombie.'
> Javier paused. 'What?'
> 'Cannibal. Ate her own grandmother.'
> Javier frowned at Amy. 'Is that true?'
> 'All the graphene. All the memory. Every last drop,' Rick said.
> Amy blinked. 'She . . . she was fighting with my mom . . . ' (Ashby 2012: ebook location 1165)

vN assumes a good level of scientific literacy on the part of its readers: 'graphene' (a form of carbon sheet one atom thick) is a term familiar to nanoscale scientists and those interested in nanotechnology but does not have widespread usage in everyday language. Biology also features in the narrative:

'You're not helping.' She rolled over onto her stomach. 'I mean, shouldn't my repair mods have rejected your stemware? I just adopted photosynthesis like . . . like a virus or something.' 'It is a virus. My pigment cells are programmed to simulate the activity of cyanophages in ocean algae. Maybe that includes turning hostiles into friendlies.' (ibid.: location 909)

Similarly with the term 'vN': John von Neumann (1903–57) was a physicist and mathematician who made important contributions to game theory and artificial intelligence theory; his name will be very familiar to physicists, in particular. Namechecking famous scientists is common in hard science fiction. Margaret Atwood's *Oryx and Crake* (2003), an account of a near-future society ravaged by a human-made bioengineering apocalypse, sent Crake, the bioengineer responsible for the disaster, to a scientific university called the Watson-Crick Institute, an homage to two of the discoverers of the structure of DNA:

> At the entranceway was a bronzed statue of the Institute's mascot, the spoat/gider – one of the first successful splices, done in Montreal at the turn of the century, goat crossed with spider to produce high-tensile spider silk filaments in the milk. The main application nowadays was bulletproof vests. The CorpSeCorps swore by the stuff. (Atwood 2003: 234)

The assumption on the part of these authors is that readers will be familiar with the language and theories being used and the scientists named, and if not will be prepared to find out more about nanotechnology and genetic engineering. The point here is that hard science fiction sees science as important to understand: it takes on a didactic role.

Hard science fiction relies on scientism for its narratives to achieve their plausibility. Throughout the subgenre there is a constant assumption that science is progressive (in that it moves forwards – perhaps with negative consequences), generates technology and will continue to do so without end. This assumption is a reminder of the lay public's attitude towards science: we may not understand it, and possibly even fear it, but we know that it will produce material change through generating new technology. No explanation in hard science fiction need be provided for plots that describe the development of nanotechnology, genetic modification, automation and robotics, spaceflight and interstellar travel. Hard science fiction sees all of these things coming about because science is working towards solving these puzzles. Indeed, hard science fiction can lay claim to having made accurate predictions of a range of future technologies. Arthur C. Clarke predicted, and was involved in, the invention of orbiting communication satellites a decade before Sputnik, the first man-made satellite, was launched, and followed that up with further predictions concerning spaceflight, personal computing and transport systems. Clarke's *Profiles of the Future* was first published in 1962, and remained in print for many years, with revisions in 1973 and 1984 (Clarke [1962] 1973). Clarke produced a new collection of future predictions with *1984, Spring: A Choice of Futures* (Clarke 1984). Other science–science fiction crossovers include what we can call 'the science of science fiction' books, didactic texts that inform readers about science by analysing science fiction narratives from a formal science perspective to determine how likely and/or plausible the stories and technologies included are. The first of these, *The Physics of Star Trek* by Lawrence Krauss (1995), uses *Star Trek* as a starting point for explaining physics to a lay audience, and includes a foreword by Stephen Hawking. Former NASA astrophysicist Jeanne Cavelos wrote a similar book using *Star Wars* as the starting point (Cavelos 1998). *The Science of Discworld* (Pratchett et al. 1999) uses the perspective of the Discworld alternative universe to investigate the planet 'Roundworld' (the Earth) and the science that explains it. For that book fantasy author Terry Pratchett was joined by mathematician and popular science writer Ian Stewart and biologist Jack Cohen. They followed it up in 2002 with a second volume, *The Science of Discworld II: The Globe* (2002), then *The Science of Discworld III: Darwin's Watch* (2005) and *The Science of Discworld IV: Judgment Day* (2013). In similar vein, sf author Stephen Baxter wrote *The Science of Avatar* (2012), a 'unique look at the science behind the blockbuster movie'; in it he notes that '[t]here is a scientific rationale for much of what we saw on the screen' (Baxter 2012: prologue). With a slightly more generic focus James Kakalios's *The Physics of Superheroes* (2010) . . . well, you've got the idea by now.

Conclusion

Science fiction is a complex and contradictory genre that takes on many forms and appeals to a large audience. We can see that the themes and ideas of science fiction are spread throughout contemporary culture; the ideas of esoteric science fiction thought communities emerge into the exoteric thought community that we inhabit and provide us with ideas and themes for interpreting the world. But the process working in the opposite direction is, perhaps, of more interest to us here. The exoteric themes and ideas of what science is, based on the scientism that dominates our considerations of knowledge in society, becomes a core element in most science fiction, providing plotlines, characters and technological possibilities for texts. The sense of scientism inside science fiction is strong, and is rarely challenged (because it is, after all, *science* fiction). This lack of a challenge to the dominant story of science we tell ourselves, and lack of challenge to the idea of inevitable progress and, often, technological determinism, is, perhaps, surprising because science fiction prides itself on being radical, having an alternative vision and taking on shibboleths and myths. This is exemplified in the way that science fiction has addressed gender and sexuality. The generally held position is that science fiction is androcentric and reproduces gender stereotypes, and there is something in this if we consider the 'golden age' science fiction of the 1940s and 1950s. But the increase in women science fiction writers and fans through the 1960s and 1970s brought about dramatic changes:

> What the feminist intervention of the 1970s did effect, though, was a new reflexivity about the conventions of SF, exposing how a genre that praised itself for its limitless imagination and its power to refuse norms had largely reproduced 'patriarchal attitudes' without questioning them for much of its existence. (Luckhurst 2005: 181–2)

We can see this in our examples of science fiction novels in this chapter. The main character in Leckie's space opera *Ancillary Justice* is Breq, formerly the spaceship *Justice of Toren*, an artificial intelligence inhabiting a female body but living in a world which doesn't care about gender. When Breq visits different cultures she finds it difficult to use the correct gender pronouns and to identify people according to gender. Throughout the book it is unclear whether her companion Seivarden is male or female; Breq always describes Seivarden using 'she' and 'her', but other characters don't. Whilst the main point of the book is not to challenge gender stereotypes, it does make the reader aware of how gender is assigned in our culture, and also to consider how gender is assigned and described in other science fiction texts. Similarly, Ashby's 'heroine' in *vN* is really three generations of 'Portias'; female humaniform robots, two generations of which inhabit one body. The main protagonist, Amy, fights external battles with human enemies and an internal battle with her grandmother, who has colonized part of her memory. Again, the main narrative of the book is not about gender identity, but Ashby is clearly making reference to the debates about gender identities and cyborg bodies initiated by Donna Haraway in the 1990s: 'A cyborg is a cybernetic organism, a hybrid of machine and organism, a creature of social reality as well as a creature of fiction' (Haraway 1991: 149).

Haraway's cyborg theory is a critique of the idea that we can find 'nature': we humans are so reliant on machinery and our bodies so invaded by technology that we have already become cyborgs. This presents possibilities for women to break away from patriarchal gender stereotyping and oppression. Haraway explicitly praises science fiction writers for generating a new myth about gender identity and boundaries that may affect how we imagine ourselves (1991: 173). For Haraway, escaping oppression means it is necessary to construct new languages, law and mythology through challenging the gender bias in existing language, law and mythology. But unlike science fiction, Haraway takes on the androcentric bias of formal science and scientism. She clearly challenges the gender bias in scientific discourse, and constructs a new mythology, one that is not dependent upon the past or transcendental, essentialist categories, in her construction of the cyborg. For Haraway, a way of re-appropriating and reconstructing science such that it loses its androcentric character and oppressive practices is to begin to understand that human beings no longer exist, having been replaced in technoscience by cyborgs. Science fiction, it seems, can only go so far along the route that Haraway maps out; it cannot challenge formal science itself. Even Margaret Atwood's *Oryx and Crake* (2003) presents science in terms of our standard account – and with a male voice.

Science fiction is reliant on science for plots and ideas, and we can track changes in these as the cutting edge of science has shifted (Marvel comic character Hulk's genesis changing from being the result of a nuclear accident in the 1960s to genetic engineering in the 2000s is a good example of this). But science fiction is not simply a sponge that picks up the themes of scientific research. Science fiction presents a version of science to audiences that will concur, at least to some extent, with audiences' hopes, and fears, for the future. The space opera vision of limitless, technologically driven human expansion and the cyberpunk vision of a dystopic, technocratic and soulless future both emerge from societies with ambiguous attitudes towards scientists, scientific institutions and scientific knowledge. Even the hard science fiction narratives of avowed cheerleaders for science as activity, and as project, see, at the very least, challenges emerging from the continuation of scientific research into areas such as nanotechnology, genetic engineering and quantum physics. In hard science fiction texts we not only learn about science as activity and as attitude, but also discover that scientific progress may be harmful to humanity – although this harm almost invariably comes from the 'corruption' of science by politics or big business. Science fiction shows us that we have a complex and contested relationship with science – a mixture of positive, negative and unknown. Science fiction tends to reconfirm existing notions of the project of science and does not disabuse us of our distorted knowledge of what actually takes place inside scientific workplaces. The science of science fiction remains, even in hard science fiction, an occluded set of processes, a black box, out of which emerge technologies, sometimes dangerous, that transform our world, scientists who retain control of crucial knowledge, institutions that exclude most members of society, and a project that has a momentum of its own and an inexorable drive towards 'progress'.

This version of science is not simply a product of the content of science fiction narratives. As we have seen, such narratives rely to a large extent on our pre-existing knowledge of science and of science fiction and its conventions. Much more important is the context within which these narratives are located: it is the context of the narrative that is providing the meaning of science that we see in science fiction. Our narratives of futurity have changed because our social understanding of the future has changed, not because we decided to construct new stories of space exploration.

Further reading

It is difficult to find definitive guides to science fiction as it is such a diverse and extensive genre. However, these two books provide a good overview of science fiction and a good starting point for further investigation:

Luckhurst, R. (2005) *Science fiction*, Cambridge: Polity.
Roberts, A. (2006) *Science fiction*, London: Routledge.

8

Science in a Changing World

Μῆνιν ἄειδε, θεά
The anger sing, goddess

Homer, *Iliad* 1.1 (author's translation)

The story of science presented in this book begins to introduce the complexity and extent of science in our technoscientific society. This story sees 'science' as a family-resemblance concept that takes on a range of meanings in different locations, but sees these meanings surrounding science as coalescing around a discourse and understanding of science that we can call scientism. Different pictures of science emerge in different places. We have seen that different thought communities will articulate discourses of science in different ways, but these discourses exist in a relationship with one another. The members of small, specialized, esoteric thought communities articulate an understanding of science that emerges from their practice and describes their work, but this discourse includes elements from the larger, exoteric thought communities of which they are also a part. Elements of esoteric discourses are, in return, incorporated into larger exoteric discourses. Identifying these pictures and places is an important task for the social understanding of science and its role in contemporary society. As we do this it becomes clear that the pictures, although all different, are all interconnected and contribute to a collective social construction of science in society. It also becomes clear that at the heart of science, at the heart of what we think of as science, there is no essence. Different pictures of science exist interdependently, but this interdependence is contingent rather than being absolute.

Looking at science in a small number of places – laboratories, popular science, science fiction, social science – is a starting point, but by no means an end point. Science is increasingly significant in all aspects of our society and culture. It infiltrates our selves and our ways of being. Given the level of institutional investment in science, both financially and ideologically, this state of affairs is very likely to continue and to intensify. Similarly, given the level of personal investment that members of society have in science and technology it is unlikely that science will dissipate and reduce in significance. Science will continue to penetrate further and further into our bodies and our minds, and the scientific will take on greater significance in global competitions for economic and military power. Yet this will not be a simple and straightforward process. Apart from the obvious point that nation states and multinational corporations will compete against each other and will attempt to control and dominate the production of crucial scientific knowledge, there is the problem of widespread public uncertainty and concern about science. Living in a culture where we are repeatedly reminded that scientific knowledge is better than other forms of knowledge, that science is external, neutral and separate from society, and that scientists have special tools and methods that allow them privileged access to the truth has some significant consequences. Nowhere is this more palpable than in societal and political responses to climate change.

Climate change, science and society

Climate change, often referred to as global warming (see box 8.1), is a major problem facing the world today. Some see it as an existential threat potentially leading to an extinction event (Lynas 2008), while others are more circumspect (Giddens 2011; Urry 2011), but all commentators who think that climate change/global warming is taking place agree that it presents major challenges to human societies.

Box 8.1 What is climate change?

The *Intergovernmental Panel on Climate Change* (IPCC) definition of climate change is as follows:

> Climate change refers to a change in the state of the climate that can be identified (e.g., by using statistical tests) by changes in the mean and/ or the variability of its properties, and that persists for an extended period, typically decades or longer. Climate change may be due to natural internal processes or external forcings such as modulations of the solar cycles, volcanic eruptions, and persistent anthropogenic changes in the composition of the atmosphere or in land use. Note that the Framework Convention on Climate Change (UNFCCC), in

its Article 1, defines climate change as: 'a change of climate which is attributed directly or indirectly to human activity that alters the composition of the global atmosphere and which is in addition to natural climate variability observed over comparable time periods.' The UNFCCC thus makes a distinction between climate change attributable to human activities altering the atmospheric composition, and climate variability attributable to natural causes. (IPCC 2014: 4)

It is only because of formal science that we know about climate change. Without this scientific knowledge we would simply experience weather, and, probably, more and more extreme weather events as global warming proceeds apace. At a certain point, unchecked global warming could render large parts of the planet uninhabitable through either sea-level rises or desertification. Without formal science we would be unaware of why this was happening, although we would still experience the phenomena. Formal science brings us the concept of climate to explain weather, analyses the atmosphere and the behaviour of the gases that comprise it, and provides us with methods for measurement of global temperature: this is how we know that CO_2 is a greenhouse gas (GHG) and is contributing to the global rise in temperature. We know from chemistry that CO_2 is released when carbon-based fuels such as coal, oil, gas and wood are burned, and we know from analysis of ice cores and other sources that the concentration of CO_2 in the atmosphere has increased continually since the industrial revolution.

There is an obvious solution to global warming; cut emissions of CO_2 by restricting or stopping the use of fossil fuels and restricting other GHG emissions that come from human activities (livestock production is a significant source of methane, a GHG that is more potent than CO_2). This is a straightforward message: fewer emissions will prevent, or at least contain, climate change. Formal science presents further clear messages: that it is likely that with continued CO_2 and other GHG emissions climate change will, at some point in the future, become runaway and catastrophic; it is already looking as if the temperature rises we will face in the twenty-first century, even the most conservative estimate of 2°C, will be extremely challenging for humanity. Temperature rises of up to 6°C have been predicted by some climate scientists: life on Earth would be almost untenable.

Observations made over decades revealed a connection between CO_2 levels in the atmosphere and temperature. As these formal science findings emerged over the years, sufficient public disquiet was generated to instigate national and international research initiatives and responses. Most significant of these is the IPCC (see box 8.2). At the same time it was becoming increasingly clear that climate change is fundamentally a social and economic problem, a consequence of

Box 8.2 The IPCC history and structure

The Intergovernmental Panel on Climate Change (IPCC) is the international body for assessing the science related to climate change. The IPCC was set up in 1988 by the World Meteorological Organization (WMO) and United Nations Environment Programme (UNEP) to provide policymakers with regular assessments of the scientific basis of climate change, its impacts and future risks, and options for adaptation and mitigation.

IPCC assessments provide a scientific basis for governments at all levels to develop climate-related policies, and they underlie negotiations at the UN Climate Conference – the United Nations Framework Convention on Climate Change (UNFCCC). The assessments are policy-relevant but not policy-prescriptive: they may present projections of future climate change based on different scenarios and the risks that climate change poses and discuss the implications of response options, but they do not tell policymakers what actions to take. . . .

IPCC assessments are written by hundreds of leading scientists who volunteer their time and expertise as Coordinating Lead Authors and Lead Authors of the reports. They enlist hundreds of other experts as Contributing Authors to provide complementary expertise in specific areas.

IPCC reports undergo multiple rounds of drafting and review to ensure they are comprehensive and objective and produced in an open and transparent way. Thousands of other experts contribute to the reports by acting as reviewers, ensuring the reports reflect the full range of views in the scientific community. Teams of Review Editors provide a thorough monitoring mechanism for making sure that review comments are addressed.

The IPCC works by assessing published literature. It does not conduct its own scientific research. For all findings, author teams use defined language to characterize their degree of certainty in assessment conclusions. IPCC assessments point to areas of well-established knowledge and of evolving understanding, as well as where multiple perspectives exist in the literature.

The authors producing the reports are currently grouped in three working groups – Working Group I: the Physical Science Basis; Working Group II: Impacts, Adaptation and Vulnerability; and Working Group III: Mitigation of Climate Change – and the Task Force on National Greenhouse Gas Inventories (TFI). (IPCC 2013)

industrial activities and modes of social and economic organization, a problem with moral and ethical dimensions and implications.

By presenting climate change as a problem that is understood through the simple device of measuring global temperature, identifying the problem as being a 0.74°C rise in global temperature in the past 100 years, and extrapolating into various futures based on modelling different risen temperature scenarios, we have produced a narrative that requires the world to restrict global temperature rise to 2°C or less (this is the agreed international, European Union and UK target). This narrative admits of only two possible solutions. The first, Plan A, is top-down multilateralism orchestrated by the UN. This plan requires individual nation states to comply with international directives to cap and then cut GHG emissions according to a strict timetable. Negotiations concerning the amount emissions should be cut by and the timetable for achieving this are long, contentious and boycotted by some nation states. Inevitably, targets are low, following lobbying by those governments and interest groups that rely on high carbon consumption and/or production. Not inevitably, but in most cases, targets are not met by individual nation states. Since the establishment of the IPCC in 1988 global CO_2 emissions have risen from about 20bn tonnes per year to 40bn tonnes per year – a 100 per cent rise. In this period there have been numerous international meetings to ensure that nation states comply with GHG reduction targets. The UNFCCC was negotiated at the Rio Earth Summit in 1992 and came into force in 1994. It contained neither guidelines for how GHG emissions should be stabilized (the goal of the treaty), nor any way of enforcing targets. It did, however, set the framework for targets to be negotiated. The Kyoto Protocol of 1997 required all nation states that were signatories to reduce their GHG emissions, and this came into force in 2005. This was followed by the Cancún Agreement of 2010, which stated that future global warming should be limited to 2°C. Given the inexorable rise in GHG emissions it is difficult to describe the UNFCCC – the backbone of Plan A – as a success.

What is Plan B? The idea that the planet can be 'fixed', 'altered' or 'saved' by technology is quite an old one. In the 1960s the Soviet Union used 'peaceful nuclear explosions' (PNEs) to create canals using H-bombs, substantially altering landscapes by this technology. The USA was also considering PNEs for similar purposes (and there are even advocates of PNEs today). Both nations came to an agreement in 1976 to limit the size of weapon yield in PNEs to 150 kilotons. Later in the Cold War the USA 'Star Wars' programme was intended to provide a space shield to protect the USA from Soviet nuclear missiles. As the existential threat of nuclear war receded (although it has still not gone away) the Star Wars programme was reconsidered to provide protection against a new existential threat: asteroid collision (Mellor 2007). Like the original Star Wars programme, the asteroid shield was never fully developed. But these programmes show that there is at least some propensity in advanced industrial societies to 'think big' when it comes to major problems. Climate change has recently been treated to this style of thinking:

'Most nations now recognise the need to shift to a low-carbon economy, and nothing should divert us from the main priority of reducing global greenhouse gas emissions. But if such reductions achieve too little, too late, there will surely be pressure to consider a "plan B" – to seek ways to counteract the climatic effects of greenhouse gas emissions by "geoengineering".' Lord Rees of Ludlow (Martin Rees) OM, President of the Royal Society. (Royal Society 2009: v)

If neither global diplomacy nor changes in human behaviour can solve climate change, a new argument is now being advanced: the world needs a Plan B for regulating climate and, more importantly, science and technology can deliver one. By intervening directly in the heat flows from the sun to the Earth's lower atmosphere, it is deemed possible by some that a thermostat for the planet could be created. (Hulme 2014: xi)

There are two main large-scale technological strategies for addressing climate change (Royal Society 2009: ix). The first involves carbon dioxide removal (CDR) techniques which remove CO_2 from the atmosphere, either by using 'synthetic trees' (an idea that appears in a number of contemporary science fiction texts already – for example, Ken MacLeod's *Intrusion* (2012)) or seeding the oceans with genetically engineered algae that will absorb CO_2, or by fertilizing the oceans with iron to stimulate phytoplankton that will extract CO_2 from the atmosphere (this was a commercial proposition from the US-based company Planktos, but their plans never got off the ground). The second set of methods involve stopping some sunlight from reaching the lower atmosphere of our planet: solar radiation management (SRM) techniques that reflect a small percentage of the Sun's light and heat back into space. These methods include building a giant mirror in space, generating clouds over the oceans (thus increasing the reflectiveness of the planet), and spraying sulphate particles into the upper atmosphere. (A prototype for the last of these, the SPICE (Stratospheric Particle Injection for Climate Engineering) project, was funded to development stage by the UK's EPSRC (http://gow.epsrc.ac.uk/NGBOViewGrant.aspx?GrantRef=EP/I01473X/1), but due to public concern and a small media storm it was cancelled in May 2012.) The aerosol injection plan has very strong support from parts of UK formal science. The SPICE project was a direct response to the Royal Society's report *Geoengineering the Climate* (2009).

The point here is not whether these geoengineering experiments will work or not. Rather it is that we look to science to fix what is, in essence, a social and economic problem, and we expect science to be able to come up with a solution. In addition, the climate change science community, perhaps reluctantly, is prepared to go along with this, to produce plans and experiments that could alter the Earth's atmosphere very dramatically. As Hulme notes:

[T]o deliberately change the condition of the planet's atmosphere in pursuit of the 'public good', in order to compensate for an induced planetary heating, . . . suggests a supreme confidence in human knowledge and ingenuity – a confidence approaching arrogance. Even to research the technologies for doing so reveals a certain poverty of the imagination, a preference for technical calculus that has little regard for the relational, creative and spiritual dimensions of *anthropos*, the human being. (Hulme 2014: 114)

Hulme is not the only commentator to consider geoengineering projects to be dangerous. Many inside the climate change science community concur, once again illustrating that formal science continues through argumentation and disagreement rather than being the large consensus that is the scientistic representation we are given. The Planktos plan to seed the ocean with iron failed due to lack of funds, although there was a serious prospect it would go ahead. When it did finally fold, *Nature* reported: 'The bad news for the company, which was unable to raise required funds, is *welcome news* for many in *the scientific community* who had been calling for a halt in such plans for years' (Courtland 2008; my emphasis – and note the collectivization of formal scientists into 'the scientific community' here). In contrast, the halting of the UK government-funded SPICE project was greeted with dismay by the president of the Royal Society, Sir Paul Nurse (Hulme 2014: 58). Eminent formal scientists and diplomats have more recently come together to advocate, as a matter of urgency, climate engineering research using, to start with, small-scale experiments. 'We must start now: gaining a solid understanding of any climate-engineering technique will take decades' (Long et al. 2015: 30).

These are contestable projects. However, media reporting of geoengineering projects is often uncritical and unreflective:

Despite the extreme nature of the proposals, journalists often fail to seek out authoritative voices (such as those of climate scientists) who are critical of such schemes. They also eschew the use of humour – despite this being a common device in UK newspapers for signalling scepticism. Most telling is the use of metaphor. Journalists rely on metaphorical systems coined by the scientists behind the proposals, rather than alternative – but equally valid – metaphorical systems with more negative connotations. Metaphors of natural processes and domestic technologies draw extreme ideas about untested global technologies into the realm of the normal and familiar. This metaphorical work renders the extraordinary ordinary and the impossible possible. Appeals to technological salvation and fantasies of control get re-configured as believable and achievable – as supposedly clean, neat alternatives to the difficult political and social changes that we must otherwise face up to. (Mellor 2008: 12)

But why has Plan A failed? Unsurprisingly, many commentators reach for a 'scientific' explanation of why people will not change their lifestyles or are in denial about climate change. Evolutionary psychology is frequently used to explain social inaction: our brains are hard-wired so that we come to ignore climate change (e.g. Marshall 2014). Social science commentators, in contrast, offer a wider range of explanations: a lack of political will on the part of governments (Giddens 2011), a lack of control of capitalism (Storm 2009), poor international institutional frameworks (Harris 2013), rampant individualism, consumerism and greed (Bauman 2013), misaligned belief- and value-systems (Hulme 2009), embedded social practices that we can't easily relinquish (Urry 2011). More sophisticated, and plausible, social science explanations are based on empirical studies and look for multiple factors that bring about a denial of climate change. Kari Norgaard's study of a Norwegian ski resort's residents' responses to climate change (the snow their local economy relied on was three months late in arriving, due to climate change) is a good example (Norgaard 2011). She found that a combination of social factors, including how people were socialized, connected to political economic discourse and overarching dominant ideas about what could and should be done. The residents not only internalized these, but also combined them with their own emotional responses to climate change. Far from not knowing about climate change, people knew a great deal. What they lacked was an appropriate forum or social space within which to speak about and share their attitudes and feelings about climate change. It was a topic whose denial was socially organized.

It is likely that Norgaard's findings are applicable to other industrial societies: we know, but we don't know how to connect the abstract science discourse of climate change to our everyday lives. When we do, through, for example, local recycling schemes, the actions we take seem minor, trivial even, in comparison to the cataclysmic discourses of global warming and climate change. However, the dominant story in our society concerning climate change is that social inaction is simply a consequence of lack of knowledge: a standard PUS model (see box 7.2 on p. 203) is applied to the problem and a standard response – giving more knowledge to people – is provided. But this cannot be sufficient; apart from anything else, members of the climate science community who clearly have all the necessary knowledge at hand and the cognitive skill to process it still make lifestyle decisions that suggest they have missed the point. In a telling, if unsystematic, piece of research, climate change activist George Marshall asked climate change experts about their travel habits:

My participants so far include a senior advisor to a leading UK climate policy expert who flies regularly to South America ('my offsets help set a price in the carbon market'), a member of the British Antarctic Survey who makes several long-haul skiing trips a year ('my job is stressful'), a national media environment correspondent who took his family to Sri Lanka ('I can't see much hope') and a Greenpeace climate campaigner just back from scuba diving in the Pacific ('it was a great trip!'). (Marshall 2009)

It is clearly unreasonable to claim that a lack of understanding of climate change science is the primary cause of a failure to change one's behaviour.

How have we got to this point? Plan A has (probably) failed and Plan B is (possibly) dangerous. But there is an abundance of knowledge about climate change, and climate change science is receiving good amounts of funding and institutional support. Most people in our society will be aware of 'global warming', and some will have changed their lifestyles as a response. But there is still a strong sense of inaction, inertia and denial. Why is this? It is reasonable to say that the story of global warming and climate change has been ignored by large sections of the populations of advanced industrial society and has even been challenged by a vociferous grouping of sectional interests who deny climate change is taking place. It is as if this scientific knowledge is particularly unwelcome and we would rather ignore it or challenge its veracity than actually act on it. To understand this we need to use sociology of knowledge.

Most scientific knowledge that emerges from esoteric scientific communities has an institutional pathway to follow that will take it from its site of production to its legitimate and institutional 'home', where it will be assimilated or used. That 'home' may be a journal or library where the knowledge will sit and wait for someone to come along and use it, or it may be a medical or corporate institutional partner 'home' where the knowledge will be assimilated into a technology transfer programme or pharmaceutical production programme (for example). Climate change scientific knowledge of global warming is different in that it is compiled from a range of formal science disciplines using disparate data sources and data points. It is constructed from elements that are self-sufficient in their own right and are usually channelled through the esoteric science communities' routes of production and assimilation. Global warming as a narrative and as a message is compiled from bringing these disparate elements together like a jigsaw puzzle and making a much bigger 'picture'. The pieces of this picture are the findings of individual esoteric thought communities, but the overall picture resides only in an exoteric context: it reflects the thought styles of quite disparate esoteric communities, but also the thought style of the exoteric community it is aiming to inform. The meaning of the formal scientific knowledge that is making up the big picture of climate change alters considerably on this passage from esoteric thought community – its site of production – to exoteric thought community (the climate change science thought community) – its site of assimilation. Not only does the scientific knowledge change its meaning, but the formal scientists change their role – from esoteric data collectors and analysers to exoteric interpreters and, in some cases, to legislators. These exoteric roles are not easy to adopt for people trained in very precise methods and users of very specialized language: problems are almost inevitable in these transitions, and we can see this clearly in terms of how climate change is interpreted, described and communicated by the climate change science community.

There is a problem with language and climate change. Wittgenstein's work shows us that meanings of words deployed in everyday language come from their use, and they are understood in context. Consider the climate change science metaphors that are regularly deployed. Moving from the esoteric discourse of data points and experimental protocols, climate change scientists translate formal science into more easily consumed forms by deploying metaphors. 'Greenhouse effect', 'global warming' and 'climate change' are the three most often encountered in public discussions about CO_2 pollution and the planet's future (interestingly, we rarely use the word 'pollution' to describe CO_2 emissions, although that is what they are). Each brings with it a range of connotations that will vary according to the context it is deployed in and the experiences of those who are hearing and using the words. None of these three common phrases is adequate to the task of explaining what is happening to the Earth's atmosphere; each causes us to interpret the 'science' in certain ways. 'Global warming' makes us think that the problem is related to the mean temperature of the planet and if we could reduce this then we would fix the problem (hence the promotion of geoengineering solutions). But this is only a very small part of the story: the bigger issue is that local weather patterns are likely to become disrupted and more extreme: adaptation to this may be a more pressing need. 'The greenhouse effect' is an interesting metaphor, both describing an actual physical process of the absorption of thermal radiation by atmospheric gases, and also taking us into the mundane and domestic realm of our everyday lives. Climate change science describes itself as 'climate change science', which seems more precise, but on reflection is, perhaps, unhelpful given that *all* climates change. But when it is added to the first term, when we put 'global warming' and 'climate change science' together, we are concatenating two terms that have emerged from formal science and which together make a very strong prediction of what is the subject of inquiry and concern (global climate), what is important (temperature) and what is happening (change through warming). When global temperatures do not rise, or when personal experiences of weather do not square with these predictions, then climate change science itself begins to look odd, and climate change deniers, who are perhaps more adept at rhetoric than formal science, leap on these apparent contradictions. Different metaphors and exoteric summaries could lead to different outcomes, or could perhaps present fewer hostages to fortune.

As an aside, it is worth thinking about the imagery we use to visualize climate change. Type 'climate change' into an online search engine and the first image that comes up is of a polar bear clinging onto the small remnants of an ice sheet. As Julie Doyle notes in her comprehensive study of how climate change is represented in the media (Doyle 2011), the fact that the polar bear has become the 'poster child' of climate change is unhelpful, but telling. Our focus on polar bears tells us that climate change is remote, is not about humans, is part of a wider and more diffuse 'save the environment' campaign that we have become inured to across the decades. The polar bear takes climate change away from us. A picture of a child sitting on the roof of a flooded house, or elderly people being rescued from flooded homes in Somerset by boats, could make us think of climate change in a much more immediate and connected way.

Further, scientism tells us a story of science as external, neutral, objective and truthful, but when climate change science is put under the intense scrutiny of the media spotlight and the forensic deconstruction that politicians and select committees will deploy, it struggles to maintain the image expected of it. Climate change science relies to a large extent on probabilities, but the story of science we tell ourselves is that science provides certainties. The disparity between these two narratives – the esoteric narrative of likelihoods, probabilities and error bars on the one hand, and the exoteric expectation of a story of truth and absolutes on the other – is difficult to bridge, particularly when science communication is so heavily distorted by expected tropes and stereotypes. Climate change science just doesn't fit the same kind of narrative structure that, say, the discovery of the Higgs boson at CERN does. There is no central locus of production, but instead a distributed network of scientists from a wide range of disciplines working on different puzzles in their own esoteric communities. Bringing these esoteric communities together to construct a global picture of climate change is a difficult and complicated matter, and one where the agonistics of each esoteric community become amplified through collectivization. IPCC reports are inevitably a compromise between the many different competing voices inside the exoteric climate change science community, even though they express a strong thought style of that community. Societal expectations of climate change science are conflicted: on the one hand we want definite statements about what is definitely going to happen and how we should deal with it – something that the climate change science community cannot and will not provide; on the other hand, we don't really want to hear the underlying message – that GHG emissions are the cause of climate change and global warming – as it may have a major impact on our lifestyles and economies. Unlike almost every other case of knowledge from inside the formal science community moving into the exoteric thought communities of our society, there is a direct consequence and a direct need for specific action to be taken whilst at the same time there is no institutional mechanism that can easily assimilate the knowledge and turn it into policies, directives or guidelines. The result is a confused and conflicted set of responses. In politics there is inertia and denial. In culture there is hyperbole and negativity (e.g. the disaster film *The Day After Tomorrow* (Roland Emmerich, 2004)). In society we find awareness and denial co-existing side by side, even in the communities most affected by climate change (e.g. Norgaard's study of a ski resort (2011)). In the opposite direction, the esoteric thought communities that make up the climate change science community are bringing into their world the thought style of their exoteric communities, working on projects that can address the 'global thermostat'.

Towards the future

The problem here lies not with the production of formal scientific knowledge, although the often poor levels of public communication that professional scientists allow to emanate from scientific institutions should be noted. The problem lies in the contestable nature of the concept of science, and the denial of this contestability by the dominant culture of scientism. We are presented with a unified and essential understanding of science, yet even a cursory glance inside scientific institutions shows this to be an inaccurate image. We are further presented with a construction of scientists as a unified and separate community that is, in some way, external to the rest of society; again quite misleading. This picture of science as unified, superior and separate is inappropriate and inaccurate. Public understanding of science rests upon these constructions, but because it is an inaccurate picture the result is a distorted and ambivalent public understanding of science. The essentialist account of science reinforces the esoteric aspects of what science is at the expense of the exoteric. This manifests itself in a number of ways.

At a very general level this can be seen in the hierarchical model of scientific knowledge, which perceives scientific knowledge and modes of explanation as being superior to other forms. The systematic exclusion of non-scientific methods from formal procedures for making sense of the world can be, and is, challenged. The institutions of formal science can also be challenged and opened up. Centralized funding of science is not necessarily the best way of paying for scientific research:

the wider public could have much more input, perhaps even at a local level, into what science is being done, where, for what reason and by whom.

However, still at a general level, the direction and scope of the project of science need to be challenged. Feminist epistemology and environmental campaigners have begun to do this, although much more movement is needed here since science, and the vested interests that surround it, are intensely conservative and resistant to change. By adopting an anti-essentialist model it is possible to see that thought communities external to science have a significant input into what science is done, and why it is done. This recognition provides a starting point for a radical reappraisal of what science is for. It should not be the case that a small range of thought communities external to science (generally elite groupings) have some input; why not all thought communities that want to?

The debate here is not about the value of science per se: few would argue that science is a 'bad thing'. Rather we need inclusive debates to take place around how science is being applied, who is doing it, in whose interests it is taking place and what the outcomes of scientific activity are. Climate change is a good example of this. In our society there are a number of competing discourses addressing climate change, emanating from a number of institutions. However, in almost all cases these institutions represent either sectional interests or closed thought communities. Thought communities collide in such debates. Formal science thought communities generate knowledge which is contestable and contested. This knowledge is translated into the understanding of what climate change is (or is not) by competing thought communities; this understanding will be based on the context, interests, knowledge, experience and thought style of the community. It will also be informed by our conception of what science is and what the status of scientific knowledge is. Promoting a dominant picture of science as the province of experts, bounded by institutional barriers that are difficult to cross, leaves the wider public in a situation where knowledge of climate change is hard to appraise: it is taken out of context by multinational corporations and governments following agendas set by financial imperatives, and on the other hand is deployed by campaign groups for specific purposes (and in some cases also for financial interests). Such co-option of knowledge makes public debate difficult. Different starting points will provide different imperatives for a community, and different understandings of what science is and what it needs to be. Yet the dominant image that receives most validation in public debates is one that excludes these other starting points.

Public debates that don't start from the position that science is external to us, that scientific knowledge is always superior and that professional scientists present neutral perspectives, would allow a wider range of voices to be heard, and would also allow all participants to see much more clearly where forces external to formal science, most notably those of business and the state, are impelling science and scientists towards particular projects and particular outcomes. Such debates might also be more acceptable to the general public in that people may be able to see that the voices emanating from their communities could be heard and may even make a difference.

The realization that scientific knowledge was produced socially, and that there was a connection between the world external to the science lab and the knowledge that emerged from it, was a huge breakthrough for social science. Sociologists, in particular, saw that to understand science it was necessary to understand society, and the relationship between science and society. However, that is only half of the story. We live with science: science surrounds us, invades our lives and alters our perspective on the world. We see things from a scientific perspective in that we will use science to help us make sense of the world – regardless of whether or not that is an appropriate thing to do – and to legitimize the picture of the world that results from such investigations. Our lives are dependent on technological devices that have emerged, in part, as a result of scientific endeavour. Every day sees yet more aspects of everyday life being scientized, be that the food we eat, the work regimes that order our labour or the medicines we take to promote health. Whilst it is important to understand the social aspects of science, the reverse may be even more important: that to understand society and culture we need to understand science.

Further reading

The IPCC collates and disseminates the most significant climate change science research, producing regular reports (the most recent at the time of writing being the Fifth Assessment Report (AR5)). These reports are very extensive, but always come with short executive summaries. The IPCC also produces regular smaller reports on studies into the effects of climate change, and climate change mitigation efforts. All IPCC publications are easily accessible through their website at http://www.ipcc.ch.

References

Agar, J. (2012) *Science in the twentieth century and beyond*, Cambridge: Polity.

Agassi, J. (1971) *Faraday as a natural philosopher*, Chicago: University of Chicago Press.

Al-Gazali, L., Valian, V., Barres, B., Wu, L.-A., Andrei, E.Y., Handelsman, J., Moss-Racusin, C. and Husu, L. (2013) 'Laboratory life: scientists of the world speak up for equality', *Nature*, 495, 7439, 35–8.

Alkon, P.K. (2002) *Science fiction before 1900: imagination discovers technology*, New York: Twayne.

Ana, J., Koehlmoos, T., Smith, R. and Yan, L.L. (2013) 'Research misconduct in low- and middle-income countries', *PLOS Medicine*, 10, 3, e1001315.

Anderson, B. (1991) *Imagined communities: reflections on the origins and spread of nationalism*, London: Verso.

Ashby, M. (2012) *vN: the first machine dynasty*, ebook, Oxford: Angry Robot.

Asimov, I. (1979a) *Foundation and empire*, South Yarmouth: J. Curley.

Asimov, I. (1979b) *Second Foundation*, South Yarmouth: J. Curley.

Asimov, I. (1982) *Foundation's edge*, Garden City: Doubleday.

Asimov, I. (1983a) *Foundation*, New York: Ballantine Books.

Asimov, I. (1983b) *The robots of dawn*, Garden City: Doubleday.

Asimov, I. (1985) *Robots and empire*, Garden City: Doubleday.

Asimov, I. (1986) *Foundation and Earth*, Garden City: Doubleday.

Asimov, I. (1988) *Prelude to Foundation*, London: Grafton Books.

Asimov, I. (1989) *Asimov's chronology of science and discovery*, New York: Harper & Row.

Asimov, I. (1993) *Forward the Foundation*, New York: Doubleday.

Atwood, M. (2003) *Oryx and Crake*, London: Bloomsbury.

Atwood, M. (2011) *In other worlds: science fiction and the human imagination*, London: Virago.

Aubrey, J. (1982) *Aubrey's Brief lives*, Harmondsworth: Penguin.

Ayer, A.J. (1971) *Language, truth and logic*, Harmondsworth: Penguin.

Baldamus, W. (1976) *The structure of sociological inference*, London: Martin Robertson.

Baldamus, W. (1977) 'Ludwig Fleck and the development of the sociology of science', in Gleichmann, P.R., Goudsblom, J. and Korte, H. (eds.) *Human figurations: essays for Norbert Elias*, Amsterdam: Stichting Amsterdams Sociologisch Tijdschrift. pp. 135–56.

Baldamus, W. (1979) 'Das exoterische Paradox der Wissenschaftsforschung: ein Beitrag zur Wissenschaftstheorie Ludwik Flecks', *Zeitschrift für allgemeine Wissenschaftstheorie*, 10, 2, 213–33. [Trans. and repr. in Erickson, M. and Turner, C. (eds.) (2010) *The sociology of Wilhelm Baldamus: paradox and inference*. Farnham: Ashgate.]

Baldamus, W. (2010a) 'Networks', in Erickson, M. and Turner, C. (eds.) *The sociology of Wilhelm Baldamus: paradox and inference*, Farnham: Ashgate. pp. 107–21.

Baldamus, W. [1979] (2010b) 'The exoteric paradox: a contribution to Ludwik Fleck's theory of science', in Erickson, M. and Turner, C. (eds.) *The sociology of Wilhelm Baldamus: paradox and inference*, Farnham: Ashgate. pp. 87–106.

Banks, I.M. (1987) *Consider Phlebas*, London: Orbit.

Banks, I.M. (1988) *The player of games*, London: Orbit.

Banks, I.M. (1990) *Use of weapons*, London: Orbit.

Banks, I.M. (1991) *The state of the art*, London: Orbit.

Banks, I.M. (1996) *Excession*, London: Orbit.

Banks, I.M. (1998) *Inversions*, London: Orbit.

Banks, I.M. (2000) *Look to windward*, London: Orbit.

Banks, I.M. (2004) *The Algebraist*, London: Orbit.

Banks, I.M. (2008) *Matter*, New York: Orbit.

Banks, I.M. (2010) *Surface detail*, London: Orbit.

Banks, I.M. (2012) *The hydrogen sonata*, London: Orbit.

Barber, B. and Hirsch, W. (1962) *The sociology of science*, New York: Free Press.

Barnes, B. and Bloor, D. (1982) 'Relativism, rationalism and sociology of knowledge', in Hollis, M. and Lukes, S. (eds.) *Rationality and relativism*, Oxford: Blackwell. pp. 21–47.

Barthes, R. (1993) 'The brain of Einstein', in Barthes, R. *Mythologies*, London: Vintage. pp. 68–70.

Bauman, Z. (2001) *Community: seeking safety in an insecure world*, Cambridge: Polity.

Bauman, Z. (2013) *Does the richness of the few benefit us all?*, Cambridge: Polity.

Baxter, S. (2008) *Flood*, London: Gollancz.

Baxter, S. (2010) *Ark*, London: Gollancz.

Baxter, S. (2012) *The science of Avatar*, ebook, London: Gollancz.

Baxter, S. (2013) *Proxima*, London: Gollancz.

Becker, G.S. (1971) *The economics of discrimination*, Chicago; London: University of Chicago Press.

Becker, W.M. (2009) *The world of the cell*, London: Pearson/Benjamin Cummings.

Behe, M.J. (2003) 'Design in the details: the origin of biomolecular machines', in Campbell, J.A. and Meyer, S.C. (eds.) *Darwinism, design, and public education*, East Lansing: Michigan State University Press. pp. 287–302.

Bennett, J., Donahue, M., Schneider, N. and Voit, M. (1999) *The cosmic perspective*, Menlo Park: Addison-Wesley.

Bennett, J., Donahue, M., Schneider, N. and Voit, M. (2010) *The cosmic perspective*, 6th edn, Menlo Park: Addison-Wesley.

Berger, P.L. (1973) *The social reality of religion*, Harmondsworth: Penguin.

Berger, P.L. and Luckmann, T. (1967) *The social construction of reality: a treatise in the sociology of knowledge*, Harmondsworth: Penguin.

Blattner, F.R., Plunkett, G., Bloch, C.A., Perna, N.T., Burland, V., Riley, M., Collado-Vides, J., Glasner, J.D., Rode, C.K., Mayhew, G.F., Gregor, J., Davis, N.W., Kirkpatrick, H.A., Goeden, M.A., Rose, D.J., Mau, B. and Shao, Y. (1997) 'The complete genome sequence of *Escherichia coli* K-12', *Science*, 277, 5331, 1453–62.

Bortolotti, L. (2008) *An introduction to the philosophy of science*, Cambridge: Polity.

Bouchaud, J.-P. and Potters, M. (2003) *Theory of financial risk and derivative pricing: from statistical physics to risk management*, Cambridge; New York: Cambridge University Press.

Bould, M., Butler, A.M., Roberts, A. and Vint, S. (eds.) (2011) *The Routledge companion to science fiction*, London: Routledge.

Bradley, H. (2013) *Gender*, Cambridge: Polity.

Bradley, H., Erickson, M., Stephenson, C. and Williams, S. (2000) *Myths at work*, Cambridge: Polity.

Braidotti, R. (2006) 'Posthuman, all too human: towards a new process ontology', *Theory, Culture & Society*, 23, 7–8, 197–208.

Broks, P. (2006) *Understanding popular science*, Maidenhead: Open University Press.

Browning, D.F., Matthews, S.A., Rossiter, A.E., Sevastsyanovich, Y.R., Jeeves, M., Mason, J.L., Wells, T.J., Wardius, C.A., Knowles, T.J., Cunningham, A.F., Bavro, V.N., Overduin, M. and Henderson, I.R. (2013a) 'Mutational and topological analysis of the *Escherichia coli* BamA protein', *PLOS ONE*, 8, 12, e84512.

Browning, D.F., Wells, T.J., França, F.L.S., Morris, F.C., Sevastsyanovich, Y.R., Bryant, J.A., Johnson, M.D., Lund, P.A., Cunningham, A.F.,

Hobman, J.L., May, R.C., Webber, M.A. and Henderson, I.R. (2013b) 'Laboratory adapted *Escherichia coli* K-12 becomes a pathogen of *Caenorhabditis elegans* upon restoration of O antigen biosynthesis', *Molecular Microbiology*, 87, 5, 939–50.

Bryson, B. [2003] (2004) *A short history of nearly everything*, London: Black Swan.

Bukatman, S. (1993) *Terminal identity: the virtual subject in post-modern science fiction*, London: Duke University Press.

Butler, D. (2013) 'Mystery over obesity "fraud": researcher baffled after his results appear in bogus paper', *Nature*, 501, 7468, 470–1.

Callon, M. (1986) 'Some elements of a sociology of translation: domestication of the scallops and the fishermen of St. Brieuc Bay', in Law, J. (ed.) *Power, action and belief: a new sociology of knowledge?*, London: Routledge & Kegan Paul. pp. 196–229.

Cavallaro, D. (2000) *Cyberpunk and cyberculture: science fiction and the work of William Gibson*, London: Athlone Press.

Cavelos, J. (1998) *The science of Star Wars: an astrophysicist's independent examination of space travel, aliens, planets, and robots as portrayed in the Star Wars films and books*, New York: St Martin's Press.

Centers for Disease Control and Prevention (2013) *Antibiotic resistance threats in the United States, 2013*, Atlanta: Centers for Disease Control and Prevention. Available at *http://www.cdc.gov/drugresistance/threat-report-2013/index.html*

Chalmers, A.F. (2013) *What is this thing called science?*, Buckingham: Open University Press.

Cho, A. (2001) 'Lost and found: the Sun's missing neutrinos were there all the time', *New Scientist*, 170, 2296, 7.

Clancy, K., Nelson, R., Rutherford, J. and Hinde, K. (2014) 'Survey of Academic Field Experiences (SAFE): trainees report harassment and assault', *PLOS ONE*, 9, 7, e102172.

Clarke, A.C. (1968) *2001: a space odyssey*, London: Arrow Books.

Clarke, A.C. [1962] (1973) *Profiles of the future: an inquiry into the limits of the possible*, New York: Harper & Row.

Clarke, A.C. (1974) *Rendezvous with Rama*, London: Pan.

Clarke, A.C. (1977) *Prelude to space*, London: Sidgwick & Jackson.

Clarke, A.C. (1984) *1984, spring: a choice of futures*, London; New York: Granada.

Clarke, A.C. and Lee, G. (1991) *Rama II*, London: Orbit.

Clarke, A.C. and Lee, G. (1992) *Garden of Rama*, London: Orbit.

Clarke, A.C. and Lee, G. (1994) *Rama revealed*, London: Orbit.

Close, F.E. (2010) *Neutrino*, Oxford: Oxford University Press.

Cohen, R.S. and Schnelle, T. (eds.) (1986) *Cognition and fact: materials on Ludwik Fleck*, Dordrecht: Reidel.

Collins, H.M. (1981) 'Stages in the empirical programme of relativism', *Social Studies of Science*, 11, 3–10.

Collins, H.M. (2010) *Tacit and explicit knowledge*, Chicago; London: University of Chicago Press.

Collins, H.M. and Pinch, T.J. (1998) *The golem: what you should know about science*, Cambridge: Cambridge University Press.

Collins, H.M. and Yearley, S. (1992) 'Epistemological chicken', in Pickering, A. (ed.) *Science as practice and culture*, Chicago: University of Chicago Press. pp. 301–26.

Courtland, R. (2008) 'Planktos dead in the water', *Nature*, online, 15 February.

Couvalis, G. (1997) *The philosophy of science: science and objectivity*, London: Sage.

Cramer, K. (2003) 'Hard science fiction', in James, E. and Mendlesohn, F. (eds.) *The Cambridge companion to science fiction*, Cambridge; New York: Cambridge University Press. pp. 186–196.

Crane, D. (1972) *Invisible colleges*, Chicago: Chicago University Press.

Cyranoski, D. (2013) 'Fallout from hailed cloning paper', *Nature*, 497, 7451, 543–4.

Cyranoski, D. and Deng, B. (2015) 'Stem-cell star lands in same venture as disgraced cloner', *Nature*, online, 11 February.

Dawkins, R. (1989) *The selfish gene*, Oxford: Oxford University Press.

Dawkins, R. (2006) *The God delusion*, London: Bantam Press.

Dawkins, R. (2004) *A devil's chaplain: selected essays*, ed. L. Menon, London: Phoenix.

de Cheveigné, S. (2009) 'The career paths of women (and men) in French research', *Social Studies of Science*, 39, 1, 113–36.

Delanty, G. (2003) *Community*, London: Routledge.

Delanty, G. (2010) *Community*, London: Routledge.

Doyle, J. (2011) *Mediating climate change*, Farnham: Ashgate.

Drexler, K.E. (1990) *Engines of creation*, London: Fourth Estate.

Einstein, A. [1916] (2001) *Relativity*, London: Routledge.

Equality Challenge Unit (2012) *Equality in higher education: statistical report 2012. Part 1: staff*, London: Equality Challenge Unit.

Equality Challenge Unit (2013) *Athena Swan Charter for Women in Science: annual report 2012*, London: Equality Challenge Unit.

Erickson, M. (2002) 'Science as a vocation in the 21st century: an empirical study of science researchers', *Max Weber Studies*, 3, 1, 29–52.

Erickson, M. (2004) 'Jean-François Lyotard: narrating postmodernity', in Robbins, D. (ed.) *Jean-François Lyotard. Vol. 3*, London: Sage. pp. 293–315.

Erickson, M. (2010) 'Why should I read histories of science?', *History of the Human Sciences*, 23, 2, 68–91.

Erickson, M. (2012) 'Network as metaphor', *International Journal of Criminology & Sociological Theory*, 5, 2, 912–21.

Erickson, M., Bradley, H., Stephenson, C. and Williams, S. (2009) *Business in society: people, work and organizations*, Cambridge: Polity.

Fanelli, D. (2009) 'How many scientists fabricate and falsify research? A systematic review and meta-analysis of survey data', *PLOS ONE*, 4, 5, e5738.

Fara, P. (2004) *Pandora's breeches: women, science and power in the Enlightenment*, London: Pimlico.

Fara, P. (2005) *Scientists anonymous: great stories of women in science*, Thriplow: Wizard.

Fara, P. (2009) *Science: a four thousand year history*, Oxford: Oxford University Press.

Feyerabend, P. (1978a) *Against method*, London: Verso.

Feyerabend, P. (1978b) *Science in a free society*, London: New Left Books.

Feyerabend, P. (1988) 'Knowledge and the role of theories', *Philosophy of the Social Sciences*, 18, 157–78.

Feyerabend, P. (2011) *The tyranny of science*, Cambridge: Polity.

Feynman, R.P. (1999) *The meaning of it all*, London: Penguin.

Feynman, R.P. (2000) *The pleasure of finding things out: the best short works of Richard P. Feynman*, London: Penguin.

Fleck, L. [1935] (1979) *Genesis and development of a scientific fact*, Chicago: University of Chicago Press.

Fleck, L. [1960] (1986) 'Crisis in science', in Cohen, R.S. and Schnelle, T. (eds.) *Cognition and fact: materials on Ludwik Fleck*, Dordrecht: Reidel. pp. 153–8.

Foucault, M. (1967) *Madness and civilization: a history of insanity in the age of reason*, London: Tavistock.

Foucault, M. (1970) *The order of things: an archaeology of the human sciences*, London: Tavistock.

Foucault, M. (1971) 'Orders of discourse', *Social Science Information*, 10, 2, 7–30.

Foucault, M. (1973) *The birth of the clinic: an archaeology of medical perception*, London: Tavistock.

Friedlander, M.W. (1998) *At the fringes of science*, Boulder: Westview Press.

Fuller, S. (1997) *Science*, Buckingham: Open University Press.

Fuller, S. (2000) *The governance of science: ideology and the future of the open society*, Buckingham: Open University Press.

Fuller, S. (2000) *Thomas Kuhn: a philosophical history for our times*, Chicago: University of Chicago Press.

Fuller, S. (2003) *Kuhn vs Popper: the struggle for the soul of science*, Cambridge: Ikon Books.

Fuller, S. (2007) *Science vs. religion? Intelligent design and the problem of evolution*, Cambridge: Polity.

Galison, P.L. (1997) *Image and logic: a material culture of microphysics*, Chicago; London: University of Chicago Press.

Galison, P.L. (2003) *Einstein's clocks, Poincaré's maps: empires of time*, New York: W.W. Norton.

Gardner, M. [1952] (1957) *Fads and fallacies in the name of science*, New York: Dover.

Gibson, W. (1984) *Neuromancer*, New York: Ace Books.

Gibson, W. and Sterling, B. (1991) *The difference engine*, New York: Bantam Books.

Giddens, A. (2011) *Politics of climate change*, Cambridge: Polity.

Glausiusz, J. (2014) 'Searching chromosomes for the legacy of trauma', *Nature*, online, 11 June.

Goldacre, B. (2008) *Bad science*, London: Fourth Estate.

Gould, S.J. (2000) *Wonderful life: the Burgess Shale and the nature of history*, London: Vintage.

Gregory, J. and Miller, S. (1998) *Science in public: communication, culture, and credibility*, New York: Plenum.

Gribbin, J. (2003) *Science: a history 1543–2001*, London: Penguin.

Griffiths, A.J.F., Wessler, S.R., Lewontin, R.C. and Carroll, S.B. (2005) *Introduction to genetic analysis*, 8th edn, New York: W.H. Freeman.

Griffiths, A.J.F., Wessler, S.R., Lewontin, R.C. and Carroll, S.B. (2008) *Introduction to genetic analysis*, 9th edn, New York: W.H. Freeman.

Hacker, P.M.S. (1986) *Insight and illusion: themes in the philosophy of Wittgenstein*, Oxford: Oxford University Press.

Hacker, P.M.S. (1997) *Wittgenstein*, London: Phoenix.

Hacking, I. (1983) *Representing and intervening: introductory topics in the philosophy of natural science*, Cambridge: Cambridge University Press.

Hacking, I. (1988) 'The participant irrealist at large in the laboratory', *British Journal for the Philosophy of Science*, 39, 3, 277–94.

Hacking, I. (1999) *The social construction of what?*, Cambridge, MA: Harvard University Press.

Hagstrom, W.O. (1965) *The scientific community*, New York: Basic Books.

Hamilton, P.F. (1996) *The reality dysfunction: book one of the Night's Dawn trilogy*, London: Pan.

Hamilton, P.F. (1997) *The neutronium alchemist: book two of the Night's Dawn trilogy*, London: Pan.

Hamilton, P.F. (1999) *The naked god: book three of the Night's Dawn trilogy*, London: Macmillan.

Hamilton, P.F. (2007) *The dreaming void*, London: Macmillan.

Hamilton, P.F. (2008) *The temporal void*, London: Macmillan.

Hamilton, P.F. (2010) *The evolutionary void*, London: Macmillan.

Haraway, D. (1991) *Simians, cyborgs, and women: the re-invention of nature*, London: Free Association Books.

Haraway, D. (1997) *Modest_Witness@Second_Millennium. FemaleMan_ Meets_OncoMouse: feminism and technoscience*, New York; London: Routledge.

Harper, B. (2009) *James Cameron's Avatar: the Na'vi quest*, New York: HarperCollins.

Harré, R. (1981) *Great scientific experiments: twenty experiments that changed our view of the world*, Oxford: Oxford University Press.

Harris, P.G. (2013) *What's wrong with climate politics and how to fix it*, Cambridge: Polity.

Hawking, S.W. (1988) *A brief history of time: from the big bang to black holes*, New York: Bantam Books.

Hawking, S.W. (2002) *On the shoulders of giants: the great works of physics and astronomy*, Philadelphia; London: Running Press.

Hawking, S.W. and Mlodinow, L. (2011) *The grand design*, London: Bantam Press.

Heisenberg, W. (1958) *The physicist's conception of nature*, London: Hutchinson.

Heisenberg, W. (1959) *Physics and philosophy: the revolution in modern science*, London: Allen & Unwin.

Henry, J. (1997) *The scientific revolution and the origins of modern science*, London: Macmillan.

Hess, D.J. (1995) *Science and technology in a multicultural world: the cultural politics of facts and artifacts*, New York: Columbia University Press.

Hobbes, T. [1651] (1968) *Leviathan*, Harmondsworth: Penguin.

Horgan, J. (1996) *The end of science: facing the limits of knowledge in the twilight of the scientific age*, London: Abacus.

Horton, H.R. (2002) *Principles of biochemistry*, Upper Saddle River; London: Prentice Hall.

Horton, H.R., Moran, L.A., Scrimgeour, G., Perry, M. and Rawn, D. (2006) *Principles of biochemistry*, Upper Saddle River: Prentice Hall.

Hughes, V. (2014) 'Epigenetics: the sins of the father', *Nature*, 507, 22–4.

Hulme, M. (2009) *Why we disagree about climate change: understanding controversy, inaction and opportunity*, Cambridge; New York: Cambridge University Press.

Hulme, M. (2014) *Can science fix climate change?*, Cambridge: Polity.

Hutchinson, I. (2011) *Monopolizing knowledge*, Belmont: Fias.

Ioannidis, J.P.A. (2005) 'Why most published research findings are false', *PLOS Medicine*, 2, 8, e124, 696–701.

IPCC (2013) *IPCC factsheet: what is the IPCC?*, Geneva: IPCC.

IPCC (2014) *Climate change 2014: impacts, adaptation, and vulnerability. Part A: global and sectoral aspects. Contribution of Working Group II to the Fifth Assessment Report of the Intergovernmental Panel on Climate Change*, Cambridge: Cambridge University Press.

Jacobs, S. (1987) 'Scientific community: formulations and critique of a sociological motif', *British Journal of Sociology*, 38, 2, 266–76.

Jacobs, S. and Mooney, B. (1997) 'Sociology as a source of anomaly in Thomas Kuhn's system of science', *Philosophy of the Social Sciences*, 27, 4, 466–85.

James, E. and Mendlesohn, F. (eds.) (2003) *The Cambridge companion to science fiction*, Cambridge; New York: Cambridge University Press.

Jancovich, M. (1996) *Rational fears: American horror in the 1950s*, Manchester: Manchester University Press.

Jayawardhana, R. [2013] (2014) *The neutrino hunters: the chase for the ghost particle and the secrets of the universe*, London: Oneworld.

Johnston, K.M. (2011) *Science fiction film: a critical introduction*, Oxford; New York: Berg.

Kakalios, J. (2010) *The physics of superheroes*, London: Duckworth.

Kant, I. (1969) *Universal natural history and theory of the heavens*, trans. Munitz, M.K., Michigan: University of Michigan Press.

Keller, E.F. (2011) 'Stick to your guns, climate scientists', *New Scientist*, 2794, 22–3.

Kim, S., Malinverni, J.C., Sliz, P., Silhavy, T.J., Harrison, S.C. and Kahne, D. (2007) 'Structure and function of an essential component of the outer membrane protein assembly machine', *Science*, 317, 5840, 961–4.

Kirby, D.A. (2003) 'Scientists on the set: science consultants and the communication of science in visual fiction', *Public Understanding of Science*, 12, 261–78.

Kirby, D.A. (2013) *Lab coats in Hollywood: science, scientists, and cinema*, Cambridge, MA; London: MIT Press.

Kirkpatrick, D. (1970) *Eduardo Paolozzi*, London: Studio Vista.

Kirkup, G., Zalevski, A., Maruyama, T. and Batool, I. (2010) *Women and men in science, engineering and technology: the UK statistics guide 2010*, Bradford: UK Resource Centre for Women in Science, Engineering and Technology.

Knorr-Cetina, K. (1981) *The manufacture of knowledge: an essay on the constructivist and contextual nature of science*, Oxford; New York: Pergamon Press.

Krauss, L.M. (1995) *The physics of Star Trek*, London: HarperCollins.

Kuhn, A. (ed.) (1990) *Alien zone: cultural theory and contemporary science fiction cinema*, London: Verso.

Kuhn, T.S. [1962] (1970) *The structure of scientific revolutions*, Chicago: University of Chicago Press.

Kuhn, T.S. (1977) *The essential tension: selected studies in scientific tradition and change*, London: University of Chicago Press.

Kuhn, T.S. and Hacking, I. (2012) *The structure of scientific revolutions*, Chicago; London: University of Chicago Press.

Lanchester, J. (2010) *Whoops! Why everyone owes everyone and no one can pay*, London: Penguin.

Larivière, V., Ni, C., Gingras, Y., Cronin, B. and Sugimoto, C.R. (2013) 'Bibliometrics: global gender disparities in science', *Nature*, 504, 7479, 211–13.

Larsen, P.O. and von Ins, M. (2010) 'The rate of growth in scientific publication and the decline in coverage provided by Science Citation Index', *Scientometrics*, 84, 3, 575–603.

Latour, B. (1987) *Science in action*, Cambridge, MA: Harvard University Press.

Latour, B. (1996) *Aramis or the love of technology*, Cambridge, MA: Harvard University Press.

Latour, B. (1999) *Pandora's hope: essays on the reality of science studies*, Cambridge, MA: Harvard University Press.

Latour, B. (2005) *Reassembling the social: an introduction to actor-network-theory*, Oxford: Oxford University Press.

Latour, B. and Woolgar, S. (1979) *Laboratory life: the social construction of scientific facts*, London: Sage.

Law, J. (1986) *Power, action and belief: a new sociology of knowledge?*, London: Routledge & Kegan Paul.

Law, J. (1991) *A sociology of monsters: essays on power, technology and domination*, London: Routledge.

Law, J. (1994) *Organizing modernity*, Oxford: Blackwell.

le Page, M. (2009) 'Get real', *New Scientist*, 2726, 31.

Leckie, A. (2013) *Ancillary justice*, London: Orbit.

Lewontin, R.C. (1993) *Biology as ideology: the doctrine of DNA*, New York: Harper Perennial.

Lloyd, G.S., Niu, W., Tebbutt, J., Ebright, R.H. and Busby, S.J.W. (2002) 'Requirement for two copies of RNA polymerase alpha subunit C-terminal domain for synergistic transcription activation at complex bacterial promoters', *Genes & Development*, 16, 19, 2557–65.

Long, J.C.S., Loy, F. and Morgan, M.G. (2015) 'Start research on climate engineering', *Nature*, 518, 7537, 29–31.

Lorditch, E. (2013) 'David Hughes: zombie ant expert and science consultant for "World War Z"' *Inside Science*. Available at http://www.insidescience.org/blog/2013/06/20/david-hughes-zombie-ant-expert-and-science-consultant-world-war-z

Luckhurst, R. (2005) *Science fiction*, Cambridge: Polity.

Lynas, M. (2008) *Six degrees: our future on a hotter planet*, London: Harper Perennial.

Lyotard, J.-F. (1991) *The inhuman: reflections on time*, Cambridge: Polity.

MacKenzie, D. (1998) *Knowing machines: essays on technical change*, Cambridge, MA: MIT Press.

MacLeod, K. (2012) *Intrusion*, London: Orbit.

Macpherson, C.B. (1968) 'Introduction', in Macpherson, C.B. (ed.) *Hobbes: Leviathan*, Harmondsworth: Pelican. pp. 9–64.

Mannheim, K. [1936] (1960) *Ideology and utopia: an introduction to the sociology of knowledge*, London: Routledge & Kegan Paul.

Marshall, G. (2009) 'Comment: why people don't act on climate change', *New Scientist*, 2718.

Marshall, G. (2014) *Don't even think about it: why our brains are wired to ignore climate change*, New York: Bloomsbury.

Mellor, F. (2007) 'Colliding worlds: asteroid research and the legitimization of war in space', *Social Studies of Science*, 37, 4, 499–531.

Mellor, F. (2008) 'Smoke and mirrors and geoengineering', paper presented at the University of Brighton Social Science Forum.

Merton, R.K. [1942] (1957) *Social theory and social structure*, New York: Free Press.

Merton, R.K. (1973) *The sociology of science: theoretical and empirical investigations*, Chicago: University of Chicago Press.

Midgley, M. (2001) *Science and poetry*, London: Routledge.

Midgley, M. (2010) *The solitary self: Darwin and the selfish gene*, Durham: Acumen.

Miller, B. (2012) *It's not rocket science*, London: Sphere.

Ministry of Defence (2013) *Annual Statistical Series 1: Finance Bulletin 1.03 – Departmental Resources*, London: Ministry of Defence.

Monk, R. (1991) *Ludwig Wittgenstein: the duty of genius*, London: Vintage.

Monk, R. (2005) *How to read Wittgenstein*, London: Granta.

Moore, A. and O'Neill, K. (2000) *The league of extraordinary gentlemen*, London: Titan.

National Science Foundation National Center for Science and Engineering Statistics (2013) *Women, minorities, and persons with disabilities in science and engineering: 2013*, Special Report NSF 13-304, Arlington: National Science Foundation. Available at http://www.nsf.gov/statistics/wmpd

Nelkin, D. (1995) *Selling science: how the press covers science and technology*, New York: W.H. Freeman.

Nelkin, D. (2004) 'God talk: confusion between science and religion – posthumous essay', *Science, Technology & Human Values*, 29, 2, 139–52.

Nelkin, D. and Lindee, M.S. (1995) *The DNA mystique: the gene as cultural icon*, New York: W.H. Freeman.

Newsome, J.L. (2008) *The chemistry PhD: the impact on women's retention*, London: UK Resource Centre for Women in Science, Engineering and Technology; Royal Society of Chemistry.

Noinaj, N., Kuszak, A.J., Gumbart, J.C., Lukacik, P., Chang, H., Easley, N.C., Lithgow, T. and Buchanan, S.K. (2013) 'Structural insight into the biogenesis of β-barrel membrane proteins', *Nature*, 501, 7467, 385–90.

Norgaard, K.M. (2011) *Living in denial: climate change, emotions, and everyday life*, Cambridge, MA; London: MIT Press.

Orwell, G. (1954) *Nineteen eighty-four*, Harmondsworth: Penguin.

Park, R. (2000) *Voodoo science: the road from foolishness to fraud*, Oxford: Oxford University Press.

Park, R. (2010) *Superstition: belief in the age of science*, Princeton: Princeton University Press.

Pearson, F. (1999) *Eduardo Paolozzi*, Edinburgh: National Galleries of Scotland.

Pearson, W.G., Hollinger, V. and Gordon, J. (2010) *Queer universes: sexualities in science fiction*, Liverpool: Liverpool University Press.

Penley, C. (1989) *The future of an illusion: film, feminism and psychoanalysis*, Minneapolis: University of Minnesota Press.

Penley, C. (1997) *Nasa/Trek*, London: Verso.

Phillips, D.L. (1977) *Wittgenstein and scientific knowledge: a sociological perspective*, London: Macmillan.

Pickering, A. (1984) *Constructing quarks: a sociological history of particle physics*, Edinburgh: Edinburgh University Press.

Pigliucci, M. (2010) *Nonsense on stilts: how to tell science from bunk*, Chicago: University of Chicago Press.

Pinch, T.J. (1986) *Confronting nature: the sociology of solar-neutrino detection*, Dordrecht: Reidel.

Polanyi, M. (1958) *Personal knowledge: towards a post-critical philosophy*, London: Routledge & Kegan Paul.

Popper, K. (1945) *The open society and its enemies*, London: Routledge & Kegan Paul.

Popper, K. (1981) 'The rationality of scientific revolutions', in Hacking, I. (ed.) *Scientific revolutions*, Oxford: Oxford University Press. pp. 80–106.

Popper, K. (2002) *The logic of scientific discovery*, London: Routledge.

Pratchett, T., Stewart, I. and Cohen, J.S. (1999) *The science of Discworld*, London: Ebury.

Pratchett, T., Stewart, I. and Cohen, J.S. (2002) *The science of Discworld II: the Globe*, London: Ebury.

Pratchett, T., Cohen, J.S. and Stewart, I. (2005) *The science of Discworld III: Darwin's watch*, London: Ebury.

Pratchett, T., Stewart, I. and Cohen, J.S. (2013) *The science of Discworld IV: Judgement Day*, London: Ebury.

Principe, L. (2011) *The scientific revolution: a very short introduction*, Oxford: Oxford University Press.

Pulver, A. (2013) 'Gravity director Alfonso Cuarón says he knew of film's scientific flaws', *Guardian*, online, 21 October. Available at http://www.theguardian.com/film/2013/oct/21/gravity-alfonso-cuaron-knew-science-flaws

Putnam, H. (1977) 'Realism and reason', *Proceedings and Addresses of the American Philosophical Association*, 50, 6, 483–98.

Radnitzky, G. (1973) *Contemporary schools of metascience*, Chicago: Henry Regnery.

Raphael, D.D. (1977) *Hobbes: morals and politics*, London: Allen & Unwin.

Reynolds, A. (2012) *Blue remembered Earth*, London: Gollancz.

Reynolds, A. (2013) *On the steel breeze*, London: Gollancz.

Ridley, B.K. (2001) *On science*, London: Routledge.

Roberts, A. (2006) *Science fiction*, London: Routledge.

Rose, H. and Rose, S. (1969) *Science and society*, Harmondsworth: Penguin.

Rose, N. (2012) 'Democracy in the contemporary life sciences', *BioSocieties*, 7, 4, 459–72.

Royal Society (2009) *Geoengineering the climate: science, governance and uncertainty*, London: Royal Society.

Russell, J.F. (2013) 'If a job is worth doing, it is worth doing twice', *Nature*, 496, 7443, 7.

Sagan, C. (1996) *The demon-haunted world: science as a candle in the dark*, New York: Ballantine Books.

Samuelson, D.N. (2011) 'Hard SF', in Bould, M., Butler, A.M., Roberts, A. and Vint, S. (eds.) *The Routledge companion to science fiction*, London: Routledge. pp. 494–9.

Sardar, Z. (2000) *Thomas Kuhn and the science wars*, Cambridge: Icon Books.

Sawyer, A. (2011) 'Space opera', in Bould, M., Butler, A.M., Roberts, A. and Vint, S. (eds.) *The Routledge companion to science fiction*, London: Routledge. pp. 505–9.

Schroeder, J., Dugdale, H.L., Radersma, R., Hinsch, M., Buehler, D.M., Saul, J., Porter, L., Liker, A., De Cauwer, I., Johnson, P.J., Santure, A.W., Griffin, A.S., Bolund, E., Ross, L., Webb, T.J., Feulner, P.G.D., Winney, I., Szulkin, M., Komdeur, J., Versteegh, M.A., Hemelrijk, C.K., Svensson, E.I., Edwards, H., Karlsson, M., West, S.A., Barrett, E.L.B., Richardson, D.S., van den Brink, V., Wimpenny, J.H., Ellwood, S.A., Rees, M., Matson, K.D., Charmantier, A., dos Remedios, N., Schneider, N.A., Teplitsky, C., Laurance, W.F., Butlin, R.K. and Horrocks, N.P.C. (2013) 'Fewer invited talks by women in evolutionary biology symposia', *Journal of Evolutionary Biology*, 26, 9, 2063–9.

Seed, D. (1999) *American science fiction and the Cold War: literature and film*, Edinburgh: Edinburgh University Press.

Shelley, M. [1818] (1994) *Frankenstein or, The modern Prometheus*, London: Penguin.

Shen, H. (2013) 'Inequality quantified: mind the gender gap – despite improvements, female scientists continue to face discrimination, unequal pay and funding disparities', *Nature*, 497, 7439, 22–4.

Shermer, M. (2001) *The borderlands of science: where sense meets nonsense*, Oxford: Oxford University Press.

Shermer, M. (2012) *The believing brain: from spiritual faiths to political convictions – how we construct beliefs and reinforce them as truths*, London: Robinson.

Silver, B.L. (1998) *The ascent of science*, New York: Oxford University Press.

Simmel, G. (1963) 'How is society possible?', in Natanson, M. (ed.) *Philosophy of the social sciences: a reader*, New York: Random House. pp. 73–92.

Sismondo, S. (2004) *An introduction to science and technology studies*, Malden: Blackwell.

Sismondo, S. (2009) *An introduction to science and technology studies*, Oxford: Blackwell.

Smaglik, P. (2014) 'Entertaining science', *Nature*, 511, 7505, 113–15.

Snow, C.P. (1959) *The two cultures and the scientific revolution: the Rede Lecture, 1959*, Cambridge: Cambridge University Press.

Snow, C.P. and Collini, S. (1993) *The two cultures*, London; New York: Cambridge University Press.

Stone, K.C. and Bennett, N.J. (2009) *Determining the Frascati compliance of MOD research & development expenditure*, DASA Defence Statistics Bulletin No. 9, Defence Analytical Services and Advice.

Storm, S. (2009) 'Capitalism and climate change: can the invisible hand adjust the natural thermostat?', *Development and Change*, 40, 6, 1011–38.

Toumey, C. (2006) 'Science and democracy', *Nature Nanotechnology*, 1, 1, 6–7.

Trenn, T.J. and Merton, R.K. (1979) 'Descriptive analysis', in Trenn, T.J. and Merton, R.K. (eds.) *Genesis and development of a scientific fact*, Chicago: University of Chicago Press. pp. 154–65.

Trigg, R. (1993) *Rationality and science*, Oxford: Blackwell.

Turner, B.S. (1992) *Max Weber: from history to modernity*, London: Routledge.

UNESCO (2007) *Science, technology and gender: an international report*, Paris: UNESCO.

Urry, J. (2011) *Climate change and society*, Cambridge: Polity.

van Frassen, B. (1980) *The scientific image*, Oxford: Clarendon Press.

Vonnegut, K. (1976) *Wampeters, foma and granfalloons*, Frogmore: Granada.

Wajcman, J. (2004) *TechnoFeminism*, Cambridge: Polity.

Wajcman, J. (2007) 'From women and technology to gendered technoscience', *Information, Communication & Society*, 10, 3, 287–98.

Wallerstein, I.M. (2004) *World-systems analysis: an introduction*, Durham, NC; London: Duke University Press.

Wallerstein, I.M. (2006) *European universalism: the rhetoric of power*, New York: New Press.

Watson, J.D. (1968) *The double helix: a personal account of the discovery of the structure of DNA*, London: Weidenfeld and Nicolson.

Weber, M. [1918] (1989) 'Science as a vocation', in Lassman, P. and Velody, I. (eds.) *Max Weber's 'Science as a Vocation'*, London: Unwin Hyman. pp. 3–31.

Westfahl, G. (2003) 'Space opera', in James, E. and Mendlesohn, F.

(eds.) *The Cambridge companion to science fiction*, Cambridge; New York: Cambridge University Press. pp. 197–208.

Williams, R. (1983) *Keywords: a vocabulary of culture and society*, London: Fontana.

Wittgenstein, L. [1921] (1981) *Tractatus logico-philosophicus*, London: Routledge & Kegan Paul.

Wittgenstein, L. [1953] (1958) *Philosophical investigations*, Oxford: Blackwell.

Wittgenstein, L. (1969) *The blue and brown books*, Oxford: Blackwell.

Wittgenstein, L. (1993) *Philosophical occasions*, Indianapolis: Hackett.

Wittgenstein, L. (1998) *Culture and value: a selection from the posthumous remains*, Oxford: Blackwell.

Yong, E., Ledford, H. and Noorden, R.V. (2013) 'Research ethics: 3 ways to blow the whistle', *Nature*, 503, 7477, 454–7.

Zuckerman, H. and Cole, J.R. (1975) 'Women in American science', *Minerva*, 13, 1, 82–102.

Index